THE GEOLOGICAL SURVEY OF WYOMING

Daniel N. Miller, Jr., State Geologist

REPORT OF INVESTIGATIONS No. 23

GOLD DISTRICTS OF WYOMING

by

W. Dan Hausel

LARAMIE, WYOMING

1980

Introduction

Sylvanite Publishing and Miningbooks.com are proud to put back into print this long out of print publication. The information in this publication is still valid and informative for historians, researchers, prospectors, miners, geologists and more. Much of this information is becoming lost as more and more books are discarded from libraries or schools, as older material disintegrates, and as publications are thrown out because someone does not know their value.

Our goal is to put hundreds of these publications back into print at a reasonable cost. Many of these originals can cost hundreds of dollars which is out of the reach of many who just want the information for study. In the process of this we try and clean up the books and text best we can to produce a quality product. There are many publishers reprinting books these days but they are essentially scan shops that put them into print errors, smudges, writing in the books, and all. Very occasionally we will do the same if we cannot obtain a copy in decent condition.

If you would like to see a particular book or are and author that has written a book in the past that is now out of print in our subject field, feel free to contact us and we will see what we can do to get that publication back in print.

CONTENTS

ILLUSTRATIONS

Figures

PREFACE

In the Fall of 1977, when I first began working for the Wyoming Geological Survey, gold was selling for $135 an ounce. At that time, most economists predicted that gold would not rise to much over $150 an ounce. But with world inflation taking its own course, many investors found gold to be a hedge against inflation and began buying the precious metal in large quantities. During the first few months of 1980, gold prices skyrocketed, and on some international markets, gold was selling for nearly $900 an ounce. More recently, gold prices appear to have shakily stabilized at just over $600 an ounce.

The price of gold has stimulated interest in gold properties in Wyoming. Several exploration groups began examining portions of Wyoming for economical gold deposits over the last two summers, and hundreds of inquiries about gold properties in Wyoming were addressed to the Wyoming Geological Survey over the last several months. Because of these requests, during the 1979 - 1980 winter, I dug through the files at the Wyoming Geological Survey and compiled this publication on many of Wyoming's historic gold districts.

This publication is intended to give the reader a brief introduction to gold deposits in Wyoming. Many of the topographical and geological maps that are listed in the sections on location and accessibility can be purchased over the counter at the Wyoming Geological Survey Building in Laramie. If the reader is interested in visiting some of the gold districts, I highly recommend that he or she take along available topographic and geological maps and review the list of references listed in the back of this publication.

In many places in the text, I've used the original price quotations for gold given in the respective references and have made no attempt to update them, primarily because of the uncertainty of the date of prices. In order to modernize many of these outdated prices, estimates for the weight in ounces were made and placed in parenthesis following the prices.

W. Dan Hausel
Staff Minerals Geologist
Geological Survey of Wyoming

INTRODUCTION

The initial gold discovery in Wyoming was probably made in the early to mid 1800's by an unnamed trapper who took time to prospect the streams between the setting of his traps. Soon after his discovery, gold had been found at several additional localities in the State, but due to hostilities between whites and Indians, early development was restricted to a handful of prospectors armed with rifles, gold pans, and sluices.

In the late 1860's to early 1870's several mining districts were organized. The foremost of these districts was the South Pass-Atlantic City District located on the southeastern tip of the Wind River Range. Total production from the South Pass-Atlantic City District is not known; however, based on some estimates, as much as 325,000 ounces of gold may have been produced from this district. In addition to the South Pass-Atlantic City District, several gold camps reported significant production. These included the Douglas Creek District in the Medicine Bow Mountains, the Negro Hill District in the Black Hills region, and the Snake River placers south of Yellowstone Park. The total production from these and all other districts of the State is not known, because no authoritative records were kept.

By the turn of the century, gold production had declined and mining interests had turned to copper production. The copper boom was short-lived; after a few decades, most of the precious and base metal mine properties were idle or abandoned. Since the 1930's, very little base and precious metal production, other than iron, has been recorded in Wyoming.

Even though production of base and precious metals has been limited to a few small prospects over the last 50 years, several areas in the state had favorable geological environments that potentially could have deposited extensive, and fairly rich mineral deposits. Many of these deposits were formerly considered subeconomic, but with the increase in precious metal prices, some of these should be re-evaluated. Of potential interest are (1) the alluvial placer deposits and the associated widespread thick sequences of Cretaceous to Tertiary gold-bearing conglomerates and sandstones in the Snake River region; (2) the Tertiary boulder conglomerates in the Dickie Springs-Oregon Gulch area; (3) the Tertiary porphyry copper-molybdenum deposits in the Absaroka Mountains; (4) the Tertiary contact metasomatic deposits in the Black Hills; (5) the Precambrian quartz-pebble conglomerates in the northern Medicine Bow and Sierra Madre and possibly the northern Laramie range; and (6) the Precambrian metasedimentary and volcanogenic environments in several of Wyoming's mountain ranges (in particular the Owl Creeks, Hartville uplift, portions of the Granite Mountains, the Sierra Madre, Medicine Bows, and the southern edge of the Wind Rivers).

Every state in the west has a legend about a lost gold mine. Wyoming is no exception. Historical records tell us that a group of Swedish prospectors found a creek in the Bighorn or Bridger mountains that contained fabulous riches of gold. According to the legend, the gold was so rich and plentiful that one could simply hand-pick the metal from the ground. Unfortunately, the prospectors were killed by Indians before they could return to the mine with a company of men. Nearly every year, one or more people will research the historical records concerning this legend and attempt to retrace the trail of the prospectors. But the Lost Cabin gold mine is still only a legend.

Currently, there are no gold mines reporting production in Wyoming. However, with the recent surges in gold prices, prospecting and exploration have been increasing in Wyoming. A number of areas of potential interest lie within regions (Wilderness, Primitive, Rare II and Roadless Study Areas) where agencies of the federal government either restrict prospecting and exploration activity or completely prohibit such activities. These restricted areas will greatly reduce the number of potential gold property developments in this state.

THE SWEETWATER DISTRICT

Introduction

The Sweetwater District, located on the southeastern tip of the Wind River Range, includes the historic gold camps of Atlantic City, South Pass, Miners Delight, and Lewiston (Plate 1). Gold was initially discovered in this region along the Sweetwater River in 1842. Over the next 25 years, prospecting and development was limited to a few small placer operations because of Indian hostilities. However, during the summer of 1867, placer gold was traced upstream to what was later to become the Carissa lode near the head of Willow Creek. Discovery of the Carissa lode led to a gold rush wherein

Figure 1. Rock Creek with piles of gravel that were dredged for placer gold during 1933 - 1941. Photograph by W. D. Hausel, Aug. 1978.

three gold camps or districts were established - South Pass City, Atlantic City, and Miners Delight. By 1869, hundreds of people were prospecting and mining in the district, and as many as 2,000 people were reported living in the camps (Beeler, 1908).

In 1871, twelve stamp mills were operating in the district, but by 1872, South Pass City was nearly deserted (Beeler, 1908): in the Atlantic City area, two mines were still active, and three mines were operating in the Miners Delight region. By 1875, the district was idle and essentially abandoned.

The discovery of the Burr lode several miles to the southeast of Atlantic City led to the establishment of the Lewiston camp (Beeler, 1908). The Burr ore was reported to be very rich; however, no production records are available for the mines of the Lewiston camp.

The closure of several mines in the district, apparently was not due to the lack of mineralization, but to mine flooding problems, milling problems, mismanagement, inexperienced miners, stock frauds and litigation problems (Beeler, 1908; Bartlett and Runner, 1926; Trumbull, 1914; Armstrong, 1947).

Since the early gold rush, a few

2

attempts have been made to reopen some of the more promising mines. Gold ore has been produced sporadically since 1879, but only in limited tonnages. Probably the highlight of the post-gold-rush era occurred during 1933 to 1941, when Rock Creek was dredged for gold (Figure 1).

Although gold-bearing rock of similar tenor to that at the outcrop was reported to continue at depth, none of the mines in the district extended more than 400 feet below the surface. Essentially, only the near-surface portions of the veins have been prospected. Large areas in the southern half of the district are covered by relatively thin layers of the Tertiary White River Formation and by Quaternary gravels and sediments. These sediments cover additional potentially gold-bearing veins that could be prospected by geochemical soil sampling techniques.

Location and Accessibility

The Sweetwater District is located on the southern edge of the Wind River Mountains in Fremont County (Figure 2). Access to the district is by paved highway from Lander or from Rock Springs. From Lander, the district lies approximately 30 miles south on State Highway 28. From Rock Springs, the district is accessible by driving 90 miles north on U. S. Highway 187 to Farson, and taking State Highway 28 north from Farson.

Topographic maps of the district include the Louis Lake, Miners Delight, Gravel Spring, South Pass, Atlantic City, Radium Springs, and Lewiston Lakes quadrangles. Geological quadrangles mapped by Bayley (1965) include South Pass City, Louis Lake, Atlantic City, and Miners Delight. Prinz (1974) prepared a geochemical map of the Atlantic City region, and Bayley and others (1973) prepared a detailed publication on the geology of the district.

The district lies in the Wind River Mountains immediately southeast of Highway 28 and of U.S. Steel's Atlantic City iron mine (Figure 3). The Atlantic City mine is presently the only active mine in the district, and produces an average of five million tons of iron ore per year (Hausel and Holden, 1978).

The iron ore is extracted from Precambrian banded iron formations (taconite) by open pit mining. The taconite is a highly magnetic, black, dense, hard rock composed of alternating bands of quartz and magnetite (iron oxide).

South Pass City and Atlantic City are located to the south and southeast of the iron mine. Both are now resort towns. The Miners Delight and Lewiston gold camps are abandoned.

Geologic Setting

The Sweetwater District is underlain by Precambrian sedimentary and volcanic rocks which formed part of a widespread system of layered rocks more than 3 billion years ago. Between 2.7 and 3.3 billion years ago, these rocks were intruded by granitic batholiths, stocks, sills and dikes. Several periods of deformation can be recognized in the layered sequence. Prior to the intrusion of the granitic magmas, the early sedimentary and volcanic rocks were folded, sheared, and metamorphosed. The primary gold-quartz mineralization probably occurred during this stage. Additional folding, faulting, and contact metamorphism occurred syngenetically with the intrusion of the granitic magmas.

During the Laramide orogeny, the Precambrian rocks were uplifted, piercing the overlying Paleozoic and Mesozoic sedimentary blanket. The uplift on a reverse fault on the western flank of the range, tilted the Precambrian core eastward. Much of the Paleozoic and Mesozoic strata were removed by erosion after uplifting.

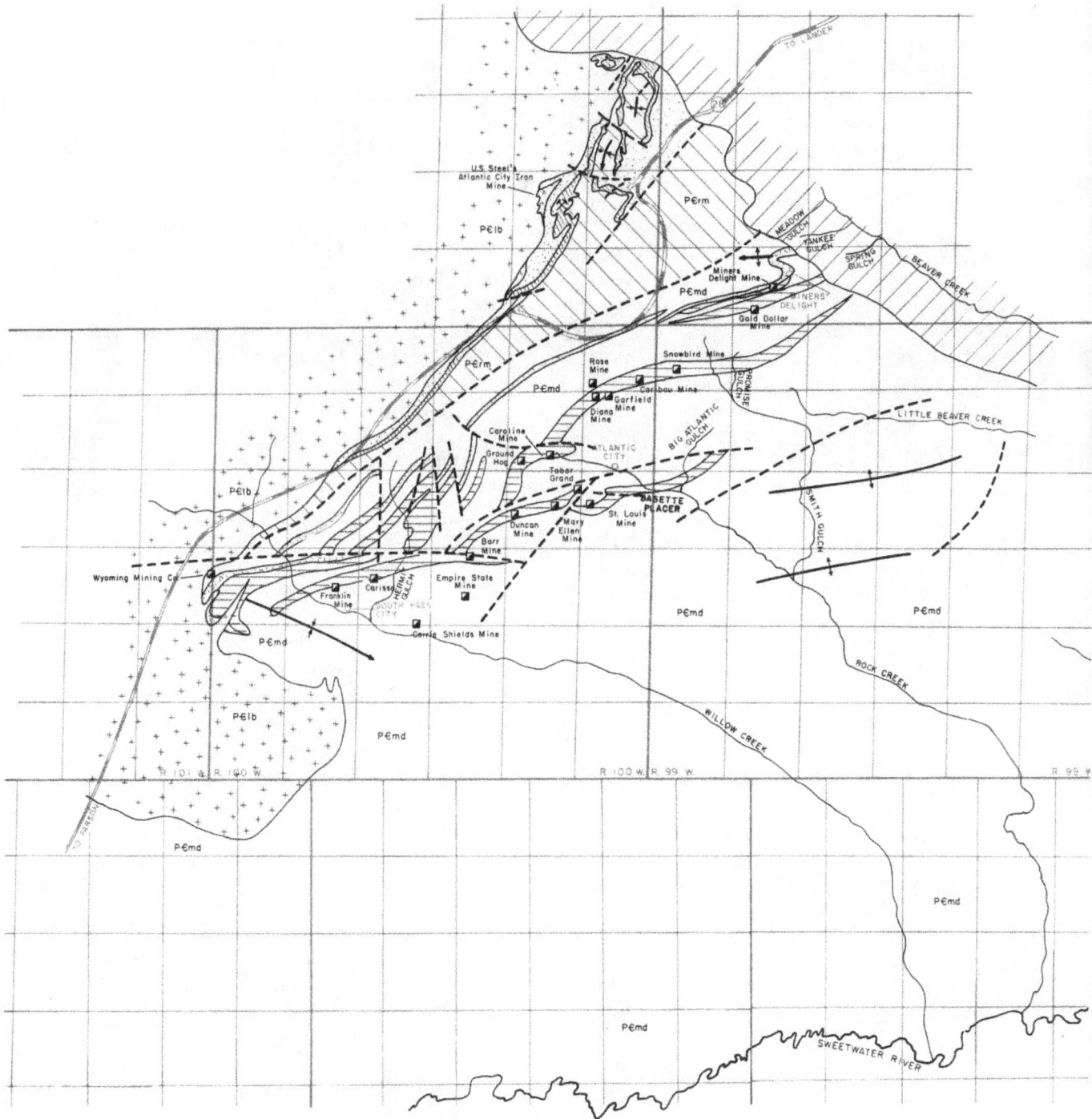

TO LANDER

U.S. Steel's
Atlantic City Iron
Mine

PЄlb

PЄrm

MEADOW GULCH

YANKEE GULCH

SPRING GULCH

BEAVER CREEK

Miners
Delight Mine

MINERS' DELIGHT

PЄmd

Gold Dollar
Mine

PЄrm

Rose
Mine

Snowbird Mine

PЄmd

Caribou Mine

Garfield
Mine

PROMISE GULCH

LITTLE BEAVER CREEK

Diana
Mine

Caroline
Mine

Ground
Hog

ATLANTIC
CITY

BIG ATLANTIC GULCH

SMITH GULCH

PЄlb

Tabor
Grand

BASETTE
PLACER

St. Louis
Mine

Duncan
Mine

Mary
Ellen
Mine

Barr
Mine

Wyoming Mining Co.

Franklin
Mine

Carissa

SOUTH PASS
CITY

Empire State
Mine

GERMAN GULCH

PЄmd

PЄmd

Gerrie Shields Mine

PЄmd

ROCK CREEK

PЄmd

WILLOW CREEK

PЄlb

PЄmd

R. 101 W. R. 100 W.

R. 100 W. R. 99 W.

R. 99 W.

TO FARSON

PЄmd

PЄmd

SWEETWATER RIVER

0   ½   1 mile

SCALE

4

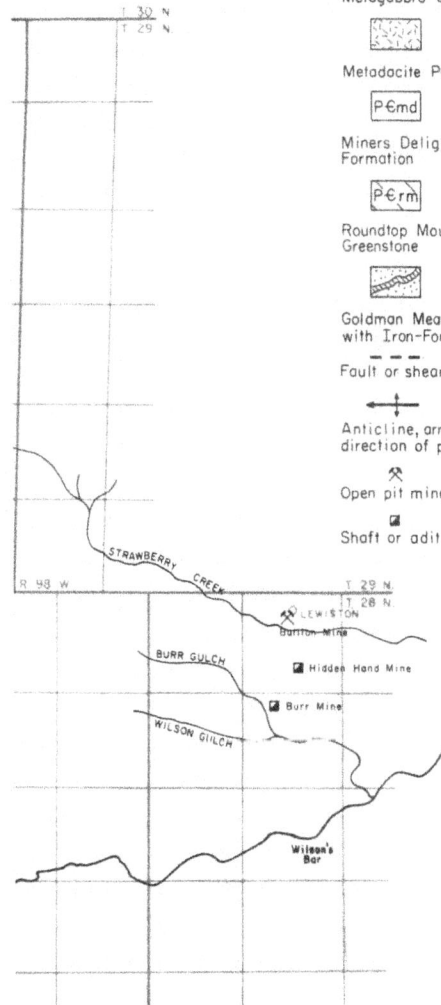

Paleozoic Sedimentary Rocks

Louis Lake Batholith (Quartz Diorite)

Serpentinite

Metagabbro & PЄmd

Metadacite Porphyry

Miners Delight Formation

Roundtop Mountain Greenstone

Goldman Meadows Formation with Iron-Formation (taconite)

- - - Fault or shear zone

Anticline, arrow indicates direction of plunge

Open pit mine

Shaft or adit

Paleozoic strata now flank the eastern edge of the district and dip gently eastward into the Wind River Basin. Later erosion stripped much of the Tertiary cover, reexposing the Precambrian basement (Bayley et al., 1973).

The majority of the district's mines lie along a northeast trending sheared metagabbro. This trend extends from South Pass City through Atlantic City and north to Miners Delight. A few scattered mines occur to the southeast of this trend, and most of these are clustered around Lewiston.

Geology

A major portion of the district is underlain by the Miners Delight Formation which is a layered sequence of metasedimentary and metavolcanic rocks deposited more than 3 billion years ago. The bulk of the formation consists of feldspathic and micaceous graywackes, conglomerates, mica schists, mafic flows, and graphic schist. During the first-stage folding of these rocks, northeast trending shear zones developed in structurally competent metagabbro and in the less competent schist and graywacke. These shears were developed conformable to the strike of bedding and folding in the schists. Silicification followed shearing and resulted in fissure filling and replacement of much of the fractured country rock. Quartz gangue was later partially replaced by arsenopyrite and gold mineralization.

Later stage faults acted as conduits for late stage quartz mineralization. These late veins contain some disseminated copper-gold mineralization but are of little economic interest (Bayley

Figure 2. Generalized geologic map of the Sweetwater District, Fremont County. Both the Quaternary and Tertiary rocks have been omitted. (Modified from Bayley et al., 1973).

5

Figure 3. View of the Atlantic City iron ore mine looking west towards the snow capped peaks of the Wind River Mountains. At present, this is the only active mine in the Sweetwater District and annually mines an average of 5 million tons of taconite ore. Photograph by W. D. Hausel, Oct., 1977.

et al., 1973). Ore shoots are commonly localized at vein intersections, or splits, or near the crest of anticlinal folds (Armstrong, 1947; Bayley et al., 1973).

Tenor of Veins

Locally, the veins contained oxidized near-surface enriched ores which were somewhat leaner at depth (Bayley et al., 1973). The tenor of the veins in the district was reported by Raymond (1870) to range from $15 to $200 (0.80 to 10.6 ounces) of gold per ton and to average about $35 (1.86 ounces) per ton. However, according to Spencer (1916), Raymond's average was probably too high, and he expected it to be in the $6 to $15 (0.32 to 0.80 ounces) per ton range. The average gold fineness was reported as 0.871 (Spencer, 1916).

Production

Gold production records are ambiguous, and in most cases nonexistant. Knight (1893) reported total production in the Sweetwater District to value $5,050,000 (260,000 ounces); Jamison (1911), Trumbull (1914), and Bartlett and Runner (1926) estimated production at more than $5,800,000 (310,000 ounces) (Table 1); whereas Spencer (1916) and De Laguna (1938) reasoned that $1.5 million and $2.0 million, respectively, were more realistic production figures.

More recently, Koschman and Bergendahl (1968) estimated gold production at 70,000 ounces.

Since 1875, only minor amounts of gold have been produced. The Duncan Mine produced nearly $40,650 (2,150 ounces) between 1911 and 1913 (Armstrong, 1947), and 890 ounces between 1946 to 1956 (notes on gold production in W.G.S. files). Additionally, the State Mine Inspector (1957) reported that 2,247 tons of gold ore were mined from the Duncan property in 1957. The Midas property produced $48,000 (1,380 ounces) in gold in 1934 (Armstrong, 1947). Between 1949 to 1954, the Carissa produced 803 ounces of gold (notes on gold production in W.G.S. files). Between 1933 to 1941, nearly $400,000 (11,500 ounces) in gold were dredged from Rock Creek (Armstrong, 1947). No additional production is reported from the district.

Sporadic panning and sluicing has continued up to the present in several of the district's streams. Spencer (1916) reported that placer gold in the district was rough, and that coarse nuggets weighing up to 2 ounces were occasionally found in the gravel.

Table 1. Estimated gold production for the South Pass - Atlantic City mines.

	Jamison's (1911) gold production estimates (in 1911 dollars)	Estimated production in ounces based on Jamison's (1911) dollar estimates	Reported production since 1911 in ounces
Miners Delight	$1,200,000	63,500	
Carissa	1,000,000	52,880	803
Caribou	500,000	26,450	
Garfield	400,000	21,150	
Victoria Regina	350,000	18,500	
Franklin	300,000	15,860	
Mary Ellen	125,000	6,600	
Lone Star	40,000	2,120	
Carrie Shields	35,000	1,850	
Ground Hog	30,000	1,585	
Young American	20,000	1,060	
Gould & Curry	20,000	1,060	
Exchange	20,000	1,060	
Doc Barr	17,000	900	
Duncan	15,000	790	3,040
Diana	10,000	530	
St. Louis	7,500	400	
Rosella	7,500	400	
Europe	7,000	370	
Charles Dickens	5,000	260	
Rose	5,000	260	
Peacock	5,000	260	
Empire State	5,000	260	
Mormon Crevice	3,000	160	
Lucky Boy	3,000	160	
Klondike	2,500	130	
Payrock	2,000	105	
Clipper	1,500	80	
Independence	1,500	80	
Meadow Gulch	1,000,000	52,880	

Table 1. Continued.

Yankee Gulch	500,000	26,450	
Spring Gulch	30,000	1,585	
Promise Gulch	30,000	1,585	
Smith Gulch	20,000	1,060	
Red Canyon	20,000	1,060	
Atlantic Gulch	15,000	790	
Beaver Creek	10,000	530	
Others	100,000	5,290	
Rock Creek			11,500
Midas Mine			1,380
Total	$5,863,000	310,050 ounces	16,723 ounces

Mine Descriptions

Carrisa Mine

location: NW¼ sec. 21, T.29N., R.100W.

production: Total value of gold production was estimated by Armstrong (1947) to be between $100,000 to $500,000 (1947 prices?). Jamison (1911), however, estimated that production from the Carissa lode was approximately $1,000,000 (52,880 ounces). In 1949, 389 ounces of additional gold were produced. During 1951, 9 ounces were produced, and in 1954, 405 ounces were produced by the Pioneer Carissa Gold Mines, Inc. (notes on gold production in W.G.S. files).

reserves: Curran (1926) reported that 75,000 tons of $7.75 ore (1926 prices) remained on the property at the time of his examination. Phillips (1911) reported 109,712 tons of $9.73 ore (1911 prices) remained.

geology: The mine is located on a shear zone trending $N70^{\circ}E$ and dipping $85^{\circ}NW$ to vertical. Mineralized quartz veins on the property occupy the shears and conform to the strike of the quartz biotite-hornblende schist country rock (Curran, 1926). Where quartz veins either split or intersect, ore shoots com-

8

monly were formed (Armstrong, 1947). The quartz gangue is partially replaced by arsenopyrite and some free gold. Spencer (1916) reported that small amounts of pyrrhotite, realgar, orpiment, and scorodite were also found on the property, but these were considered as gangue. The tenor of the Carissa ore averaged $6 (0.32 ounces) per ton and ranged from a trace to $50 (2.64 ounces) of gold per ton (Beeler, 1908).

The width of the vein averaged 6-1/3 feet in thickness and was developed by a 400 foot deep shaft and a total of 2300 feet of drifts and crosscuts on five levels (Bartlett and Runner, 1926). A stamp mill was constructed on the property to process the ore (Figure 4).

Figure 4. Historic South Pass City with the Carissa mine and mill in the background (University of Wyoming, Western History Research Center photograph).

Empire State Mine (B&H Mine)

    location: Sec. 22, T.29N., R.100W.

    production: Jamison (1911) estimated that $5000 (260 ounces) in gold were produced on this property.

geology: The geology at the mine is complicated by north-south and east-west faulting and it is thought that the ore vein occurs on the south limb of an anticline. There appear to be two main vein systems which conform to the strike of the fault trends. Vein thicknesses vary from a few inches up to 7 feet wide. The richest ore was found at the intersection of the main vein (strikes N5°E) and the cross vein (strikes N80°E). The north-south vein was reported to contain pyrite with arsenopyrite, whereas the east-west vein was high grade and free milling. The silver-to-gold ratio is 2 to 1, the highest reported in the district (Armstrong, 1947).

Duncan Mine

location: W½ sec. 14, T.29N., R.100W.

production: Jamison (1911) estimated that $15,000 (790 ounces) in gold were mined from the Duncan Mine. Between 1911 and 1913 an additional 2,150 ounces were produced. In 1946, 76 ounces of gold and one ounce of silver were produced (Armstrong, 1947). In 1955, 52 ounces were produced, and in 1956, 762 additional ounces of gold were extracted (notes on gold production W.G.S. files).

geology: The vein varies from 18 inches to five feet wide and is in excess of 200 feet in length. The ore vein occupies a shear zone entirely within amphibolite.

At the southeast end of the property, the main vein (strike-N65°W) is intersected by a vertical cross vein which strikes N45°E. The Duncan shaft was developed on the vein intersection to a depth of about 250 feet. Several hundred feet of drifts, crosscuts, and open stopes were developed from the shaft. The gold is reported to average 0.820 fine (Armstrong, 1947) (Figure 5).

Mary Ellen Mine

location: NE¼ sec. 14, T.29N., R.100W.

production: Jamison (1911) estimated production at $125,000 (6,600 ounces).

geology: Quartz veins cut granodiorite and trend N40°W (dip 35°SW)

10

and N40°E (dip 30°SW) on the property. The vein intersections form the ore shoot which was stoped 240 feet down dip (Armstrong, 1947). The average ore tenor was estimated at $8 (0.42 ounces) in gold per ton (Jamison, 1911).

Figure 5. The historic Duncan Mine near Atlantic City. Photograph by W. D. Hausel, Aug., 1978.

Tabor Grand Mine

location: Sec. 14, T.29N., R.100W.

geology: Two shafts were developed 180 and 160 feet deep with a few hundred feet of drifts, crosscuts and winzes. At about 120 feet below the collars the shafts were intersected by an adit driven in from the draw immediately east of the shafts. Average ore grade was reported to have been $10 (about ½ ounce) a ton. The vein strikes N85°W and dips vertically (Armstrong, 1947).

Carrie Shields Mine

location: SE¼ sec. 21, T.29N., R.100W.

production: Total estimated production for this mine prior to 1911 was $35,000 in gold (1,850 ounces) (Jamison, 1911). The average grade of the ore was estimated at nearly $20 (1 ounce) in gold per ton (Armstrong, 1947).

geology: The ore zone is 3 to 4 feet wide; it is a belt of fractures with discontinuous quartz stringers. The mineralized zone cuts gray to greenish biotite-schist that strikes N.65°E. and dips 85°S. The vein itself strikes N.65°E. and dips 65°S.

The property is developed by a 180 foot deep vertical shaft with a few hundred feet of drifts. An adit driven from the draw to the southwest intersects the shaft 65 feet above its deepest level (Armstrong, 1947).

Midas Mine

location: One mile north of Atlantic City (exact location unknown).

production: The mine is reported to have produced $48,000 (1,380 ounces) in 1934 (Armstrong, 1947).

geology: A polished section of Midas ore shows that late gold and arsenopyrite entered along fractures in the quartz (Bane, 1929).

Rock Creek Placers

location: Along Rock Creek.

production: Between 1933 and 1941, 3,000,000 cubic yards of auriferous gravel were mined which averaged $0.125 in gold per cubic yard and in some spots up to $0.50 per cubic yard (Armstrong, 1947).

geology: Seventy-five percent of the gold on Rock Creek was found within 1 to 3 feet of bedrock. No gold was found in gravel overlying schist, but only in gravels overlying blocky amphibolite. The richest gravel was located 1 mile below Atlantic City near the place where a large fault crosses Rock Creek (Armstrong, 1947). In the $S\frac{1}{2}$ secs. 20, 21, 22 and $N\frac{1}{2}$ secs. 27, 28, 29, T.30N., R.100W., the gold was reported to average $0.55/yd., with some spots assaying as high as $41.00/yard. In 1953, it was mined by the Wyoming Mica and Metals Corporation (Wilson, 1953). The gold averaged 0.880 to 0.896 fine, and usually carried one part silver to every 10 parts gold (Armstrong, 1947).

Big Nugget Claim

    location: Secs. 31, 32, 33, T.29N., R.98W. and sec. 6, T.28N., R.98W.
                    along Strawberry Creek and the Big Nugget Ditch.

    production: Five "good-sized nuggets" are reported to have come from
                    the property (Haff, 1944).

Black Rock - Long Creek Area

    location: North of the Sweetwater River, near Rongis.

    geology: The geology is similar to that of the South Pass area. Schist
                    layers in granite trend east-west, and consist of hornblende
                    schist and mica schist. The rocks contain "diorite" dikes
                    and are cut by iron-stained quartz veins. The King Solomon
                    Claim (sec. 36, T.31N., R.92W.) contained northeast trending
                    veins which were stained by limonite and malachite (Osterwald
                    et al., 1966).

Diana Mine

    location: About 1½ miles northwest of Atlantic City.

    geology: Gold is associated with quartz veins in a micaceous schist.
                    One sulfide zone was developed on shows of arsenopyrite and
                    gold. Another vein contained pyrolusite. The White Horse
                    Mining Company reported the ore to average $25 (0.72 ounces)
                    per ton in gold, with some assays up to $100 (2.88 ounces)
                    per ton (Wilson, 1951).

Ground Hog Group

    location: One-half mile north of the Duncan Mine.

    geology: A dike of "altered diorite," 20 to 100 feet wide, contains
                    several small quartz veins. The zone is reported to contain
                    $5 to $8 (0.26 - 0.42 ounces) in gold per ton (Jamison, 1911).

Lander Belle Mine

    geology: A dike of silicified quartz porphyry lying between granite
                    and schist is about 200 feet wide, and contains between
                    $2.50 and $5.00 (0.13 - 0.26 ounces) in gold per ton
                    (Jamison, 1911).

Miner's Delight Group

location: Sec. 32, T.30N., R.99W.

geology: Gold-bearing quartz veins are associated with dikes that cut schist, quartzite, and slate. The dikes and metamorphic rocks strike northeast and dip to the north. Native gold is disseminated throughout the matrix, and in vein openings. The main vein is 3 to 14 feet wide, and may be up to 2,000 feet long (Kyner, 1907). The mine is located at the east end of a schist body, just west of the sedimentary (Paleozoic?) section.

Riverton-Kinnear Area

location: Tps. 1 and 2 N., R.2E.

geology: Placer gold values at four localities in the Wind River gravels near Neble ranged from 26 to 56 cents per cubic yard. Thirteen to 28 miles upstream from Riverton, gold values in the Wind River gravels are reported to range up to 32 cents per cubic yard (Bolmer and Biggs, 1965).

Overland District

location: Located 35 miles south of Lander.

geology: Quartz veins and trachyte dikes cut basaltic dikes and other volcanic rocks all of Precambrian age. The veins contain copper sulfides, carbonates, and oxides. Assays yielded 5 to 54 percent copper, $0.85 to $975 (0.04 to 51.5 ounces) per ton of gold, variable amounts of silver, and iron varying from 2 to 27 percent. Smelter return from one shipment was reported to have averaged $20 per ton (1913 prices). The property contains shallow shafts 10 to 40 feet deep with shallow cuts on veins (Keeton, 1913).

Wyoming Copper Mining Co.

location: S½ sec. 13 and N½ sec. 24, T.29N., R.101W. Located about one mile west of South Pass (see Spencer, 1916, plate 11).

production: A quartz vein in schist contained 4 to 16 percent copper, $2 to $10 (0.11 to 0.53 ounces) in gold per ton, and 1.5

to 40 ounces of silver per ton. The vein is 40 feet wide. The property was operated in the summer of 1914 and a shaft was opened to a depth of 500 feet with some drifting. Small amounts of rich copper mineralization were found on the surface, but nothing of value was discovered at depth. The project was abandoned in the fall of 1914 (Spencer, 1916).

Garfield Mine

location: N½ sec. 12, T.29N., R.100W.

geology: The Buckeye shaft developed on the property reached a depth of 140 feet in 1870, and production at that time amounted to $50,000 (2,655 ounces) in gold per year. In 1905, 10,000 tons (264 ounces) of ore were produced, valued at $5,000 (Spencer, 1916).

Copper Surprise Mine

location: One mile north of South Pass City.

geology: The prospect was developed on a chalcopyrite vein containing gold values. A 135 foot shaft was sunk on the property prior to 1908 (Beeler, 1908).

Burr Mine

location: NW¼ sec. 8, T.28N., R.98W. in the Lewiston District.

geology: The Burr Mine was discovered in 1879 and reportedly was very rich (Beeler, 1908). No production records are available.

Sweetwater Placers

geology: Gold-bearing gravels (Figure 2) reported by Spencer (1916) include: (1) the Carissa Placer, located on the east end of South Pass City in a north-south drainage that drains from the Carissa Mine to Willow Creek; (2) Hermit Gulch, located to the east of South Pass City; (3) Babette Placer, which occurs on the south side of Rock Creek and to the southwest of Atlantic City; (4) Big Atlantic Gulch, located about one mile east of Atlantic City; (5) Smith Gulch, located about 1½ miles east of Big Atlantic Gulch; (6) Promise

Gulch, which drains into Smith Gulch northeast of Atlantic City; (7) Meadow, Yankee, and Spring gulches, which drain into Beaver Creek northeast of Miners Delight; (8) the upper portions of both Beaver and Little Beaver Creeks; (9) the Burr placer, located in the Burr drainage immediately west of the Burr Mine; (10) Strawberry Creek, south of Lewiston; (11) Willow Creek, in the vicinity of South Pass City; (12) Rock Creek, south of Atlantic City; and (13) Wilson's bar, south of Lewiston on the Sweetwater River.

Spencer (1916) reported that nuggets weighing 2 ounces were found in many of these placers.

DICKIE SPRINGS - OREGON GULCH DISTRICT

Introduction

Gold placers in the Dickie Springs - Oregon Gulch region were discovered along the historic Overland Trail in 1863. In that same year, the discoverers were exterminated by Indians. Between 1864 to 1882, the district was abandoned because of hostilities between the whites and Indians, but was again opened with the signing of the "Treaty of Five Nations" in 1882 (Greene, 1896). Following the reopening of the district, several prospectors and miners actively began mining gold from the placer alluvial deposits. The total amount of gold extracted from the district is not known. At any rate, the total gold resource contained in the alluvial gravels is significant. Greene (1896) reported that the placer gravels contained a total gold resource value of $45,662,-324 (1896 prices) (2,207,000 ounces).

Reports by both Greene (1896) and Love et al. (1978) suggest that the district contains significant amounts of gold. Two major factors which have restricted the development of gold in this district are (1) much of the gold in this region is finely disseminated throughout a boulder conglomerate facies of the Wasatch Formation; and (2) the district is barren of water except during the Spring. The closest source of water is the Sweetwater River, which lies nearly 6 miles to the north.

Location

The Dickie Springs - Oregon Gulch District lies south of the Sweetwater District near the Fremont - Sweetwater County border (Plate 1). The district is accessible from State Highway 28, which connects Farson (to the south) with Lander (to the north). From Lander, the district is reached by 41 miles of paved road and 6 miles of graded light-duty road. Topographical maps covering the district include the Pacific Springs, Dickie Springs, and Continental Peak quadrangles. Geological publications about the district include Zeller and Stephens (1964; 1969) and Love et al. (1978). The following narative is abstracted from

16

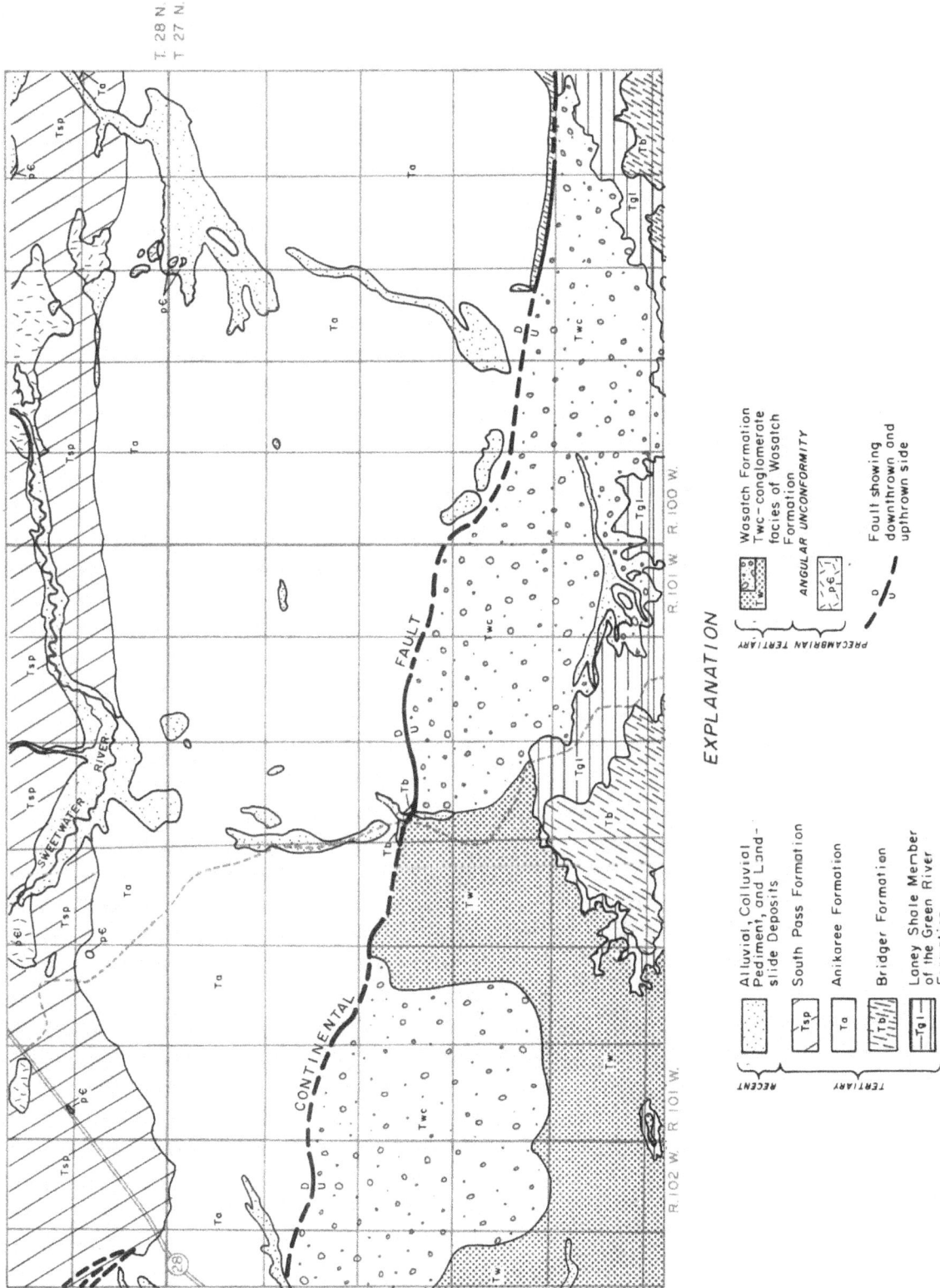

Figure 6. General geology of the Dickie Springs - Oregon Gulch District (after Love et al., 1978; Zeller and Stephens 1964).

the paper of Love et al.

Geology

The Dickie Springs - Oregon Gulch District contains rocks of Tertiary age unconformably overlying Precambrain units (Figure 6). Of primary economic interest are rocks of the Wasatch Formation and Recent alluvium eroded from the Wasatch, in that these units are significant gold bearers.

The Wasatch Formation consists of boulder conglomerate facies separated laterally by sandstone and claystone. The conglomerate facies contains giant boulders (as large as 25 feet in diameter) in a brown arkosic matrix. The boulders were derived from the Precambrian terrain of the Wind River Mountains, and compositionally range from granitic boulders to ultramafic clasts. Fragments of gneiss and schist are also common.

The conglomerate facies is divided into a western and an eastern lobe by nearly 2 miles of sandstone and siltstone units also of the Wasatch Formation. These units are of the same general stratigraphic horizon as the conglomerate facies. Evidence suggests that the conglomerate unit to the west in the Pacific Butte area was derived from a different region in the Wind River Mountains than the facies to the east of Dickie Springs. This is based on radically different gold concentrations. Geochemical and trace element studies of the Dickie Springs - Oregon Gulch gold suggest that the gold originated from hydrothermal veins in a predominately granitic terrain. This conclusion suggests that the gold was primarily derived from a region other than in the South Pass - Atlantic City area, where the gold has mafic, rather than granitic, affinity.

Economic Geology

Some of the richest gold concentrations in the Dickie Springs - Oregon Gulch district are found in alluvial deposits north of the Continental Fault. These deposits were worked by the early prospectors and were derived from the erosion of the Wasatch conglomerate facies. Gold concentrations in these alluvial deposits average about 0.01054 ounces per cubic yard of gravel (Love et al., 1978). Greene (1896) estimated that 5,843 acres of the placer deposits averaged 0.0387 ounces per cubic yard, with spots of gravel running from as high as 0.7334 ounces to as low as 0.0029 ounces per cubic yard. The gold is relatively coarse, and many flakes average 0.2 inches in diameter.

The conglomerate facies on the western end of the district average about 0.000183 ounces per cubic yard, whereas the eastern facies average 0.00262 ounces per cubic yard and range from 0.00002 to 0.10377 ounces of gold per cubic yard of gravel (Love et al., 1978). The total gold content estimate in these conglomerates is significant.

The minimum stratigraphic thickness of these conglomerates is 1,300 feet, and they have a combined areal extent of eight square miles. The total amount of gold contained in this volume of rock may exceed 28,-500,000 ounces (Love et al., 1978), at a present value of more than $17 billion (at $600 an ounce).

18

THE DOUGLAS CREEK DISTRICT

Introduction

The Douglas Creek District was established after 1868 when Iram Moore discovered gold in stream-deposited gravels in a tributary of Douglas Creek in Albany County. The initial discovery site was later named Moore's Gulch. By 1869, the district was active with many prospectors using sluices, rockers, and gold pans to concentrate and separate the precious metal from stream gravels.

During the first year of heavy prospecting (1869) about 425 ounces of gold having a value of $8,000.00 (1906 prices) were found in Douglas Creek. Beeler (1906) reported that many of the washings yielded $2 to $2.50 (0.11 to 0.13 ounces) in gold to the pan. The gold varied from fine- or flour-gold to coarse-gold with flat nuggets up to 1/8 inch in length. Nuggets weighing from 5 to 20 pennyweights, (one pennyweight is equal to one twentieth of an ounce) were common, with the largest reported nugget weighing 68 pennyweights (nearly $3\frac{1}{2}$ ounces).

The auriferous gravels lie on predominately granitic rock and range from 3 to 15 feet thick, averaging about 5 feet in thickness. Gold occurs throughout the gravels and is concentrated at the contact of the gravels with the granite bed rock. In addition to gold, platinum has also been extracted from some drainages. The platinum is more commonly found in the northern half of the district, in the vicinity of the Albany placers.

The coarseness of some of the gold suggested that it was derived from a nearby source. This led to prospecting upslope from the drainages and to the eventual discovery of lode deposits. The more important mines in the district were the Centennial, Keystone, Florence, and Rambler mines.

Some prospecting still continues in the district, but primarily by recreationists using gold pans and small one- or two-man floating dredges. Some gold is found every year on Douglas Creek.

Location and Accessibility

The Douglas Creek District lies on the eastern flank of the Snowy Range (Medicine Bow Mountains) within Tps. 13, 14, and 15 N., and Rs. 78 and 79 W. (Plate 1). The district is accessible from Laramie by State Highways 130 and 230. On these highways, the district lies to the west of Laramie. By way of Highway 230, the district is reached by traveling approximately 35 miles southwest to the Foxpark turnoff. From the Foxpark turnoff, Douglas Creek is reached by driving $1\frac{1}{2}$ miles north to Foxpark and then turning west from Foxpark. Douglas Creek lies approximately $7\frac{1}{2}$ to 8 miles from Foxpark.

By way of State Highway 130, Douglas Creek is reached by driving 25 miles west to State Highway 11. From Highway 11, the town of Albany is reached after driving 11 miles. From Albany, Douglas Creek lies 8 miles by graded dirt road. Topographic maps covering the district include the Foxpark (1961), Keystone (1961), Horatio Rock (1961), Medicine Bow Peak (1961), Overlook Hill (1961), Elkhorn Point (1961), and Albany (1961) quadrangles.

Geological maps of the district are found in Houston and others (1968, plates 1 & 4), in Currey (1965), and in McCallum and Orback (1968).

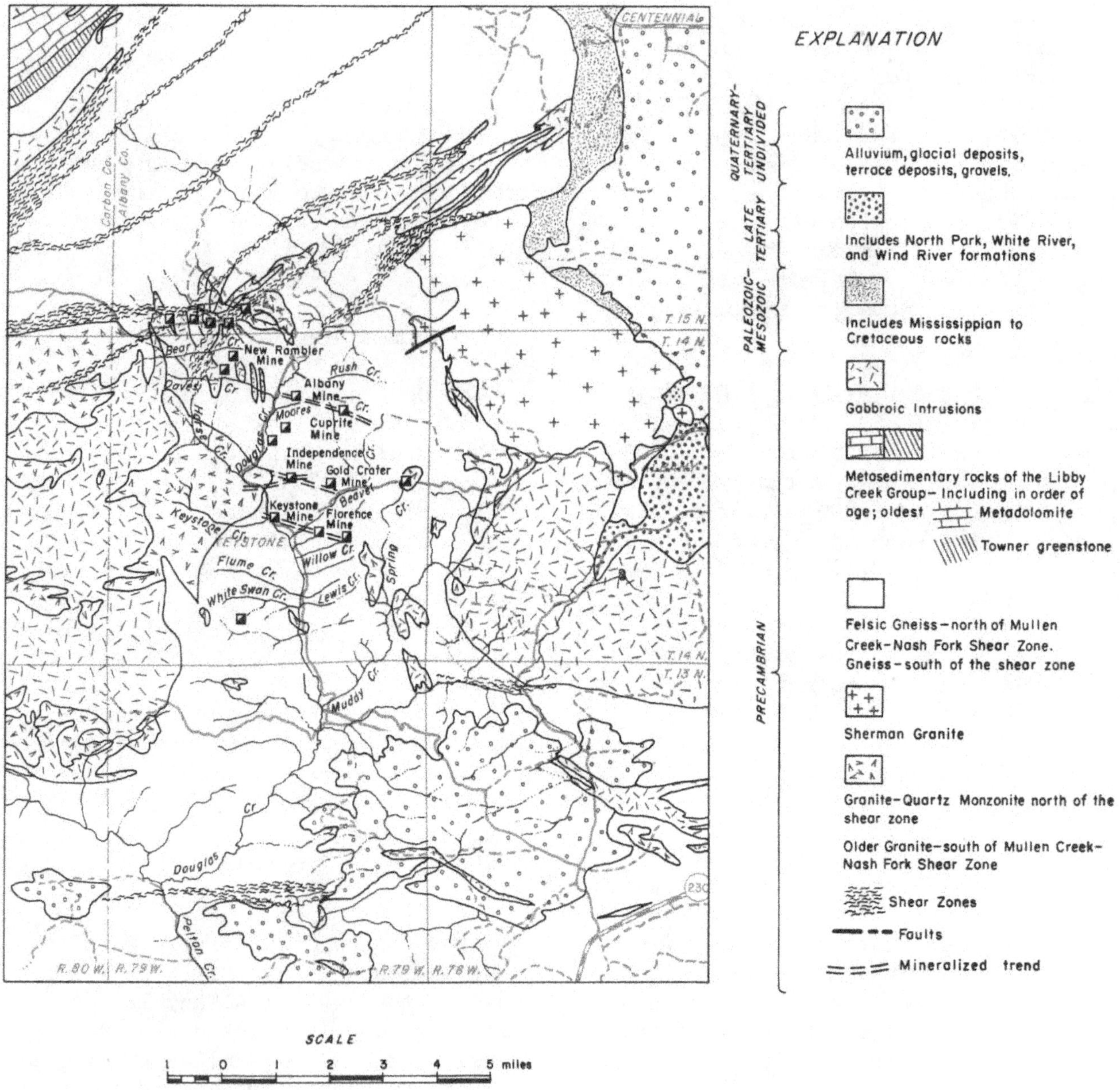
EXPLANATION
QUATERNARY-TERTIARY UNDIVIDED
Alluvium, glacial deposits, terrace deposits, gravels.
LATE TERTIARY
Includes North Park, White River, and Wind River formations
PALEOZOIC-MESOZOIC
Includes Mississippian to Cretaceous rocks
PRECAMBRIAN
Gabbroic Intrusions
Metasedimentary rocks of the Libby Creek Group- Including in order of age; oldest Metadolomite
Towner greenstone
Felsic Gneiss-north of Mullen Creek-Nash Fork Shear Zone. Gneiss-south of the shear zone
Sherman Granite
Granite-Quartz Monzonite north of the shear zone
Older Granite-south of Mullen Creek-Nash Fork Shear Zone
Shear Zones
Faults
Mineralized trend
New Rambler Mine
Albany Mine
Cuprite Mine
Independence Mine
Gold Crater Mine
Keystone Mine
Florence Mine
KEYSTONE
CENTENNIAL
Carbon Co.
Albany Co.
SCALE
1 0 1 2 3 4 5 miles

Figure 7. Generalized geologic map of the Douglas Creek District (modified from Houston et al., 1968).

Geology and Mineral Deposits

Primary mineral deposits in the Douglas Creek District occur in mineralized quartz veins occupying shear zones. In the southern half of the district (south of Rush and Dave creeks), primary mineralization is found as copper-gold quartz veins in narrow shear zones which trend northwest (N60°W; Curry, 1965). Intersecting mineralized trends are reported at (1) the Independence Mine - where an east-west trend intersects the N60°W trend, and (2) at the New Rambler mine, located in the northern half of the district (Figure 7).

In the northern half of the district, primary deposits of copper and gold associated with quartz veins trend northwest. These deposits occur in fracture fillings, in fissure fillings, and in brecciated and shattered portions of shear zone tectonites. The New Rambler mine is located at the point of intersection of northwest and northeast trending shears (McCallum and Orback, 1968).

Curry (1965) suggests that the primary mineralization in the southern half of the district (Keystone area) resulted from late-stage emplacement of a granitic melt. In the northern half of the district, mineralization is related to mafic intrusions (McCallum and Orback, 1968). Supergene enrichment was important in the formation of rich ore deposits at the New Rambler Mine. Ore deposits in the supergene zone often assayed higher than 35 percent copper with some

Table 2. Gold production in the Douglas Creek District. After Knight (1893).

	Dollar Value	Estimated Ounces
Centennial Mine	$ 50,000	2,630
Keystone Mine	100,000	5,260
Florence Mine	35,000	1,840
Other Mines	4,000	210
Placer Mines	40,000	2,100
Total	$229,000	12,040

platinum. Extensive supergene enrichment appeared to be absent at other mine localites.

The placer gold and platinum concentrations are found in stream gravels and are more highly concentrated near the gravel-bedrock contact. Placer gold was reported to occur as fine- to flour-gold with nuggets as large as 1/8 inch. Many panned nuggets weighed between 16 and 68 pennyweights (0.8 to 3.4 ounces) (Beeler, 1906).

Production

The total gold-platinum-copper production in the Douglas Creek District is not known with any great accuracy. Knight (1893) estimated that approximately $229,000 (12,040 ounces) in gold were produced in the district (this estimate excluded the New Rambler Mine). McCallum and Orback (1968) reported that the minimum copper-gold-platinum production from the New Rambler Mine was $120,000.

At today's prices, 12,040 ounces would be worth about $7,225,000 (gold at $600 an ounce).

Douglas Creek
Placer Mines

The Douglas Creek Placer Mines extended along much of Douglas Creek proper and several of its tributaries. The most important gold-bearing tributaries included Lake Creek, Muddy Creek, Spring Creek, Keystone Creek, Beaver Gulch, Horse Creek, Gold Run, Joe's Creek, Moore's Gulch, Dave's Creek, Elk Creek, Bear Creek, and Willow Creek (Figure 7). The district embraces a fifteen mile long, ten mile wide area lying 45 miles west of Laramie.

Much of the placer gold was reported to be coarse. A few nuggets contained considerable quartz, suggesting an origin from a nearby quartz vein. Nuggets weighing between 0.8 and 3.4 ounces were reported. The greater portion of the gold was in the shape of finer particles from fine- to flour-gold, although flat nuggets up to 1/8 inch were mined (Beeler, 1906).

The auriferous gravels range between 3 and 20 feet in thickness and average 5 feet thick. The greater gold concentrations were found at the gravel-bedrock contact.

Douglas Creek Consolidated Placers

location: The Douglas Creek Consolidated placers include a group of placer claims extending from sec. 10, T.13N., R.80W. to sec. 2, T.13N., R.79W., a distance of about 8 miles. And the placers extend for about 5 miles along Muddy Creek to the south line of section 18, T.14N., R.78W.

gravel tenor: Beeler (1906) ran pan tests on several claims within the Douglas Creek Consolidated placers, obtaining the following results: Near the intersection of Douglas Creek with Pelton Creek (sec. 19, T.13N., R.79W.) pan tests in a 160-foot traverse ranged from $0.90 to $1.75 (0.048 to 0.093 ounces of gold) per cubic yard and averaging $1.16 (0.061 ounces). Average thickness of these gravels was reported as 5 feet.

A 150-foot traverse along Douglas Creek north of Keystone in the SW¼SW¼ sec. 15, T.14N., R.79W. was pan tested at $0.925 (0.049 ounces) per cubic yard. About 900 feet above this traverse, a crosscut traverse averaged $1.05 (0.056 ounces) to the cubic yard. The average gravel depth is approximately 6 feet.

In sec. 25, T.14N., R.79W. on Muddy Creek, a pit measuring 48 feet by 15 feet pan tested at $1.25 (0.066 ounces) per cubic yard.

Average tests on Douglas Creek gave $0.835 (0.044 ounces) per cubic yard with a minimum value of $0.35 (0.019 ounces) per cubic yard in gold (1906 prices). Gold sent to the Denver Mint from Douglas Creek was 0.911 fine, and from Lincoln Gulch was 0.950 fine and contained some platinum (Beeler, 1906).

Home Placers

location: Located above (to the north of) the Douglas Creek Consolidated placers. These placers include secs. 10, 15, 21, 22, 27, and 34, T.14N., R.79W. on Douglas and Beaver Creeks.

gravel tenor: The largest nugget (3.4 ounces) discovered on Douglas Creek was found in the Home placer gravels. Above the mouth of Beaver Creek, the gravels contain granitic boulders 1 to 4 feet in diameter, and the drainage is 80 to 250 feet wide and not favorable to work. Below Beaver Creek, the gravel is not as coarse and the drainage opens up into Willow Flat, an open area 600 to 800 x 2000 feet with gravels varying in thickness from 3 to 8 feet. The gold averaged $0.18 to $0.24 (0.009 to 0.013 ounces) per cubic yard in Willow Flat (Beeler, 1901; 1906).

Albany Placers

location: These placers are located to the north of the Home Placers and extend up Douglas Creek proper and include Moore's Gulch, Elk Creek, Bear Creek, and Dave's Creek. The Rob

Roy Reservoir was constructed below the mouth of Moore's
Gulch on Douglas Creek proper, flooding major portions of
Moore's Gulch, Rush Creek, Dave's Creek, Bear Creek, and
Elk Creek.

gravel tenor: Beeler (1906) reported that the average gravel on Douglas
Creek within the Albany placers returned $1.25 (0.066
ounces) per cubic yard in gold for the year of 1906. Much
of the gold was coarse and jagged with considerable flour-
gold and a trace of platinum. The gold was reported as
0.890 to 0.959 fine. Other tests were run along Dave's
Gulch: in one test, twenty-two hundred cubic yards of gravel
returned about 190 ounces in gold, averaging about $1.60
(0.085 ounces) per cubic yard. These gravels also contained
metallic platinum in the black sands. Some palladium was
also reported (Beeler, 1906).

On Moore's Gulch, 60,000 cubic yards of gravel tested
at over $1.00 (0.053 ounces) per cubic yard. During 1869
and 1870, 4,000 cubic yards of gravel were mined, returning
about $9,000 (475 ounces) in gold. In many places, $0.50
to $2.50 (0.026 - 0.132 ounces) in gold were panned from
the gravels and many nuggets were found weighing from one
to 20 pennyweights.

It was estimated that Douglas Creek proper on this
property contained 3,020,160 cubic yards of gravel which
should not run less than $0.50 (0.026 ounces) per cubic
yard (Beeler, 1906). Elk Creek and Bear Gulch were estimated to contain about 250,000 cubic yards of gravel (Beeler,
1906).

Spring Creek Placers

location: The Spring Creek placers extend about 2 miles, along the
entire length of Spring Creek in secs. 24, 25, and N½ 36,
T.14N., R.79W.

gravel tenor: Spring Creek is an intermittent creek which carries suffi-
cient water for prospecting only during the Spring. In
1895, 1,200 cubic yards of gravel were produced returning
about $1,000 (50 ounces) in gold. The gold in Spring

24

Creek is coarse and jagged (about 40 percent of the 1895
gold production was in the form of nuggets weighing from
1 to 17 pennyweights) (Beeler, 1906).

Small Placers

location: Above the mouth of Muddy Creek. Approximately sec. 2,
             T.13N., R.79W.

gravel tenor: The gravel reportedly averaged $2.00 (0.11 ounces) of gold
              per cubic yard (Beeler, 1906).

Lincoln Creek Gulch

location: Lincoln Gulch extends about 3 miles in E½ sec. 5 and sec. 9,
             and 16, T.13N., R.78W.

gravel tenor: Prior to 1906, $400 to $1,600 (20 to 85 ounces) in gold was
              taken annually from the Lincoln Gulch Placers. In places,
              the auriferous gravels are 20 feet thick (Beeler, 1906).

Douglas Creek Lode Mines

After gold was discovered in the
stream deposits, it was assumed,
because much of the gold was coarse
and several nuggets contained considerable quartz, that the gold had
not traveled far from its source.
Several prospectors began searching
for quartz veins carrying gold, resulting in the discovery of shear
zone quartz veins carrying values in
gold and copper. In the southern
half of the district, in the vicinity
of the Independence mine, two mineralized trends were recognized: an
east-west shear zone termed the "Monarch" trend and a northwest trend
termed the "Mammoth" trend. These two
trends intersected at the Independence
Mine, and ore was found to be localized
at the vein intersection. Several additional mines were located on trends
parallel to the "Mammoth" trend.

In the northern half of the district, several mines were located on
northeast trending shear zones. For
example, the New Rambler lode was
localized at an intersection of a
northeast trending shear with a northwest trending shear.

Albany Mine

location: Located on a northwest mineralized trend which extends
             through Moore's Gulch in sec. 10, T.14N., R.79W.

geology: A 360-foot shaft was sunk by the year 1903 on the Albany-

Cuprite mineralized trend. At a depth of 150 feet, a covellite-bearing ore body was encountered. No gold was reported (Curry, 1965).

Blanche Mine

location: SE¼ sec. 32, T.15N., R.79W. Located immediately west of the New Rambler Mine.

geology: The Blanche Mine was sunk on a mineralized shear zone in an attempt to intersect an extension of the New Rambler ore body. The shafts are located on sheared felsic gneiss, metagabbro, and metadiorite. The main shaft was sunk to a depth of 160 feet and penetrated the ore-bearing zone at 120 feet. The ore zone contained copper carbonates, chalcocite, and chalcopyrite (Beeler, 1906). Copper mineralization was found in quartz veins associated with pyrite, hematite, and limonite, and in gouge in shear zones (Osterwald et al., 1966; McCallum and Orback, 1968).

Cuprite Mine

location: NW¼ sec. 11, T.14N., R.79W. Along the Albany-Cuprite trend.

geology: A 954-foot drift was driven off of a 65-foot deep shaft in 1900. The mineralized zone contained native copper, cuprite, pyrite, chalcopyrite, gold and silver. The vein assayed 3 to 28 percent copper, a trace to 2.56 ounces per ton of gold, and a trace to 2 ounces of silver per ton. Some colbalt and chromium were also reported (Osterwald et al., 1966; Curry, 1965).

Douglas Mine (Morning Star)

location: In the SE¼ sec. 9, T.14N., R.79W. along the western bank of Douglas Creek.

geology: A 150 foot deep shaft encountered a seven foot wide mineralized zone at the 35 foot level. Eighty feet of drifts and crosscuts were developed on the mineralized veins. The zone contained native copper, copper carbonates, chalcopyrite, chalcocite, cobaltite, and gold. Three veins, 6 inches to 2 feet, 2 to 3 feet, and one foot in thickness, were reported at deeper mine levels (Curry, 1965). The surface workings were destroyed by the construction of the road on the west bank of Douglas Creek.

Dutchess Mine

location: SW¼ sec. 32, T.15N., R.79W. West of the Blanche Mine.

geology: Several exploratory shafts and pits were sunk on strongly sheared metagabbro and metadiorite. A steam plant was constructed on the property anticipating production, but the mine apparently did not produce any ore. Traces of copper associated with pyrite, hematite, and limonite in quartz were found on the dumps (McCallum and Orback, 1968). The mineralized shear zone trends northeast.

Florence Mine

location: Near Keystone, located along the Keystone-Florence mineralized trend, which strikes northwest. The mine was developed in the SE¼ sec. 22, T.14N., R.79W.

geology: Gold occurred in auriferous pyrrhotite found in a 3 to 5 foot wide quartz vein. Ore concentration was difficult, in that much of the ore was finely divided and not "free-milling" (Beeler, 1906). The ore was not amenable to concentrating, but was considered better suited to treatment by chlorination or cyanide. Auriferous pyrrhotite kidney zones were exceedingly rich, sometimes ranging from 7.5 to more than 48 ounces per ton; however, this ore was very discontinuous. A 160-foot shaft was developed with stopes and drifts on both the 30 foot and the 100 foot level (Curry, 1965).

Gold Crater Mine

location: The Gold Crater Mine was located on the eastern extension of the Mammoth trend in NE¼ sec, 22, T.14N., R.79W.

geology: Several small quartz veins reportedly carried "free-milling" gold, pyrite and chalcopyrite. The average ore tenor was reported to be equivalent to one foot of 20 dollar ore (1905 prices) (Beeler, 1905). The mine was last worked in 1937 (Curry, 1965).

Independence Mine

location: Located at the intersection of the Mammoth and Monarch mineralized trends in sec. 15, T.14N., R.79W.

geology: In the early 1900's, an 80-foot shaft with 100 feet of crosscuts was sunk at the intersection of two mineralized veins within quartz-biotite schist country rock. A 125-foot drift was driven in along the Monarch trend about ¼ mile west of the shaft. The ore contained chalcanthite and averaged 13.5 percent copper with some reported bismuth (Curry, 1965).

Keystone Mine

location: The Keystone Mine is found on the Keystone-Florence trend in the NW¼ sec. 22, T.14N., R.79W.

geology: The vein was developed from a 365-foot shaft with about 5,000 feet of drifts. In 1890, a 40-ton stamp mill was built on the property (Figure 8). In 1893, operations ceased with 6,000 tons of ore on the dump reportedly ready to process with an addtitional 100,000 tons of identified reserves reported underground. The vein average 1.2 fine ounces of gold per ton.

The gold occurred as "free" gold and also occurred in pyrite, pyrrhotite, and in mylonite selvage adjacent to the quartz veins (Curry, 1965). The average value of the ore was $23.50 per ton, and total production was estimated at $96,000 (Osterwald et al., 1966). In the 1950's the Keystone mine plant was dismantled and the shaft sealed (Curry, 1965).

Figure 8. Historic photograph of the Keystone Mine and stamp mill about 1900 (University of Wyoming Western History Research Center photograph).

Lake Creek Mine

location: Near the junctions of Muddy and Lake creeks with Douglas Creek (SE¼ sec. 2, T.13N., R.79W).

geology: Copper-gold mineralization occurs in silicified mylonite within a broad east-west shear zone. Gold, silver, and copper assays ranged from $4.00 to $140.00 per ton. This mine was developed by a 75-foot shaft with 715 feet of crosscuts and drifts intersected by a 335-foot adit (Curry, 1965).

Fairview Claim

location: Sec. 6, T.13N., R.79W.

geology: A 3.5 foot wide vein reportedly assayed at 28.5 percent copper and 0.8 ounces of gold per ton. The assay was taken from the vein at a depth of eight feet (Osterwald et al., 1966).

Kansas Group

location: (?) sec. 12, T.13N., R.79W. Located on Lake Creek nearly 6 miles southeast of Keystone.

geology: A 40-foot shaft was developed in a 6-foot quartz vein in quartz diorite country rock with a 150-foot adit drifted to the intersection of the vein. The vein was stained by limonite and copper carbonates, and yielded some gold values (Beeler, 1907).

Maudem Group
  location: SW¼SW¼ sec. 1, T.13N., R.79W. on Lake Creek
  geology: Assays on a limonite- and copper-stained quartz vein gave considerable gold value according to Beeler (1904). Samples of chalcocite and chalcopyrite were found on the mine

a

b

Figure 9. Views of (a) the New Rambler Mine tailings and (b) mine dump (photographs by W. D. Hausel, July 1979).

30

dump (Osterwald et al., 1966). Three shafts and two adits were developed on the property.

New Rambler Mine (Holmes Mine)

location: NW¼ sec. 33, T.15N., R.79W.

geology: Production from the Rambler mine (Figure 9) totaled 6,080 tons of copper with traces of platinum, palladium, gold, and silver (Osterwald et al., 1966). The Rambler Mine was developed at the intersection of northwest and northeast trending mineralized shear zones. The near-surface oxidized ores were gold bearing. At nearly 65 feet depth, rich supergene copper ore was discovered containing platinum group metals (McCallum and Orback, 1968).

THE CENTENNIAL RIDGE DISTRICT

Introduction

The Centennial Ridge district was organized in 1876 after gold was discovered in stream gravels (unpublished report on the Centennial Mine, W.G.S. files, 1940). Several lode discoveries followed. Following the announcement of platinum discovered in the New Rambler Mine near Douglas Creek, a new wave of prospecting activity in the Centennial Ridge district was undertaken to locate the valuable precious metal.

Total production in the district is not accurately known. Dart (1929) reports that at least $90,000 in gold was mined from the Centennial mine. At least 100 tons of ore were milled from the Free Gold claim (Hess, 1925); the Utopia realized some production, but records are not available to show the amount of ore.

Location

The Centennial Ridge district lies within one mile of the town of Centennial (Plate 1). The mines are reached by Highway 130. From about 2 miles west of Centennial, across from the U.S. Forest Service Information Center, the district is reached by driving west approximately 2 miles on a graded dirt road. At the intersection with the Cliff Gold Mine trail, the district lies about one mile south.

The Centennial 1:24,000 topographic quadrangle (1961) covers the district. A geological map published in McCallum (1968) also extends over the district.

Economic Geology

Both primary and placer deposits are reported in the district (Figure

10). Primary gold deposits are found in quartz veins that parallel the foliation and schistosity of biotite and hornblende gneisses and schists. Primary gold-platinum veins occur as fracture fillings and replacements in shear zones and faults which cut the biotite and hornblende gneisses and schists. The platinum group metals are associated with sulfides and arsenides in the fracture fillings.

The richest ores in the district are associated with sulfide-rich zones in the mafic host rocks. The greatest sulfide concentrations are generally found in graphitic fault gouge, brecciated mafic mylonites, and strongly chloritized wall rock. Sulfide veins up to four inches thick have been reported in drifts cutting the upper Middle Fork Creek shear zone. Pyrite is the predominant sulfide, except where it is oxidized to limonite, although small amounts of arsenopyrite and cobaltian pyrite are sometimes present. Chlorite, epidote, calcite, dolomite, siderite, alum, and minor quartz occur with the sulfide masses. Chalcopyrite may occur locally along with small amounts of malachite, azurite, and chrysocolla (McCallum, 1968).

The gold and platinum minerals are believed to have been derived from the hydrothermal leaching of mafic rock sulfides during the Precambrian. The metal-bearing solutions would have been deposited as fracture fillings in open spaces in faults and fractures. The mafic rocks are the most probable source of the metals because of the intimate relationship of the metals with the mafic biotite and hornblende schists and gneisses (McCallum, 1968).

Placer gold and platinum occur in alluvial gravels along the Middle Fork of the Little Laramie River and along Queen Mill run and Fall Creek. No production records are available from the placer deposits. However, according to E. K. Burhans (personal comm. to McCallum), gravel reportedly carried free gold, platinum metals, galena, and free mercury (?) (McCallum, 1968).

Mine Descriptions

Centennial Mine

location: SE¼ sec. 4, T.15N., R.78W.

production: The ore at the Centennial mine was reported to average 1.5 ounces of gold per ton. Production records indicate that about $90,000 (4,780 ounces) of gold were milled in the late 1860's. The ore from the Centennial mine was reported to be rich: an ore sample from the Centennial mine won first prize at the Paris Exposition in 1878 (unpublished report on the Centennial Mine, W.G.S. files, 1940; Dart, 1929).

geology: The ore occurs as free-milling gold associated with quartz veins and shear zones in iron-stained mafic hornblende

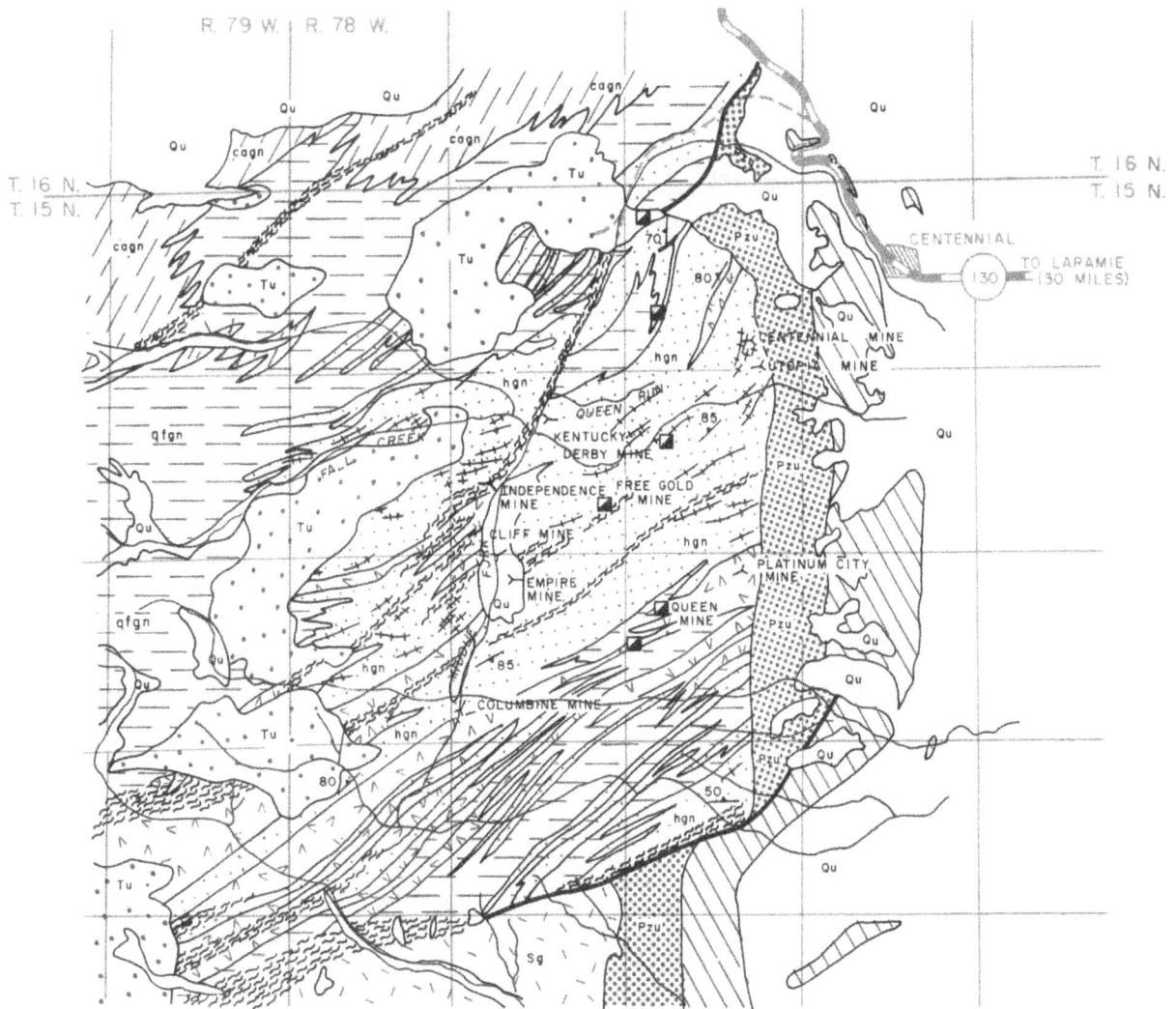

EXPLANATION

Qu	Quaternary undivided		cagn	Cataclastic biotite augen gneiss.
Tu	Tertiary sandstone, arkosic sandstone, interbedded with conglomerate		qfgn	Quartzo-feldspathic gneiss
	Mesozoic rocks undivided		hgn	Hornblende gneiss
Pzu	Paleozoic rocks undivided		85	Strike and dip of foliation
Sg	Sherman Granite			Mine adit
	Pegmatite			Mine shaft
	Mafic igneous rock Predominately gabbro, norite, pyroxenite, and amphibolite.			Fault
				Zone of intense shearing
				Drainage

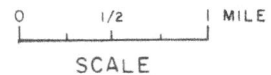

SCALE
0   1/2   1 MILE

Figure 10. Generalized geology of the Centennial Ridge gold-platinum district (after Houston et al., 1968; McCallum, 1968).

33

gneiss and schist country rock (McCallum, 1968). The
N45°E trending mineralized vein was intersected by an adit.
Mining terminated when the vein abruptly ended at a fault.
Later attempts to locate the faulted vein failed (unpub-
lished report on the Centennial mine, W.G.S. files, 1940).

Utopia Mine
    location: N½NE¼ sec. 9 and S½SE¼ sec. 4, T.15N., R.78W.
    geology: This property was developed by three adits in the early
                 1900's (Figure 11). Gold is reported associated with
                 quartz veins, shear zones, and as inclusions in garnets in
                 hornblende schist country rock (Dart, 1929). The ore
                 veins were offset by faulting which resulted in the closure
                 of the mine (Hess, 1926). Assays were reported to range
                 from a trace to 1.64 ounces of silver and a trace to 3.46
                 ounces of gold per ton (Dart, 1929).

Free Gold Claim
    location: SE¼ sec. 8, T.15N., R.78W.
    production: At least 100 tons of gold-quartz ore were milled on the
                 property (Hess, 1926).
    geology: Hess (1926) reported that the Free Gold claim was developed
                 by nearly 800 feet of surface workings along a two-foot-
                 wide quartz vein carrying free gold and oxidized pyrite.
                 The vein is enclosed by hornblende and chlorite schist
                 paralleling the schistosity.

Platinum City Mine
    location: NW¼NE¼ sec. 16, T.15N., R.78W.
    geology: The Platinum City mine was developed on pegmatite, quartz,
                 and sulfide veins in amphibolized meta-igneous rocks.
                 Quartz veins reportedly contained free gold. Sulfide
                 veins and pods commonly were gold and platinum bearing
                 (McCallum, 1968).

PLAN VIEW OF UTOPIA MINES
(WITH ASSAYS)

Utopia #4

54"-0-0.04
42"-0-0.008
48"-0-0.02
52"-0-trace
41"-0-0.05
56"-0-0.05
46"-0-0.03
53"-0-0.080
55"-0.08-0.02
104"-0-0.14
58"-0.06-0.12
60"-0-0.14
108"-0.16-0.14
54"-0-0.12

N

Utopia #3

Fault

Utopia #2

10"-0-0.52
38"-0.64-1.36
26"-0-0.58
10"-0.28-0.62
54"-0.38-0.52

10"-0-0.88
10"-0-0.66
4"-0-0.54
6"-1.64-3.46
9"-0-0.74

Fault

CROSS SECTIONAL VIEW OF
UTOPIA MINES

Utopia #2

Utopia #3

Fault

Utopia #4

Fault

0   50 feet
SCALE

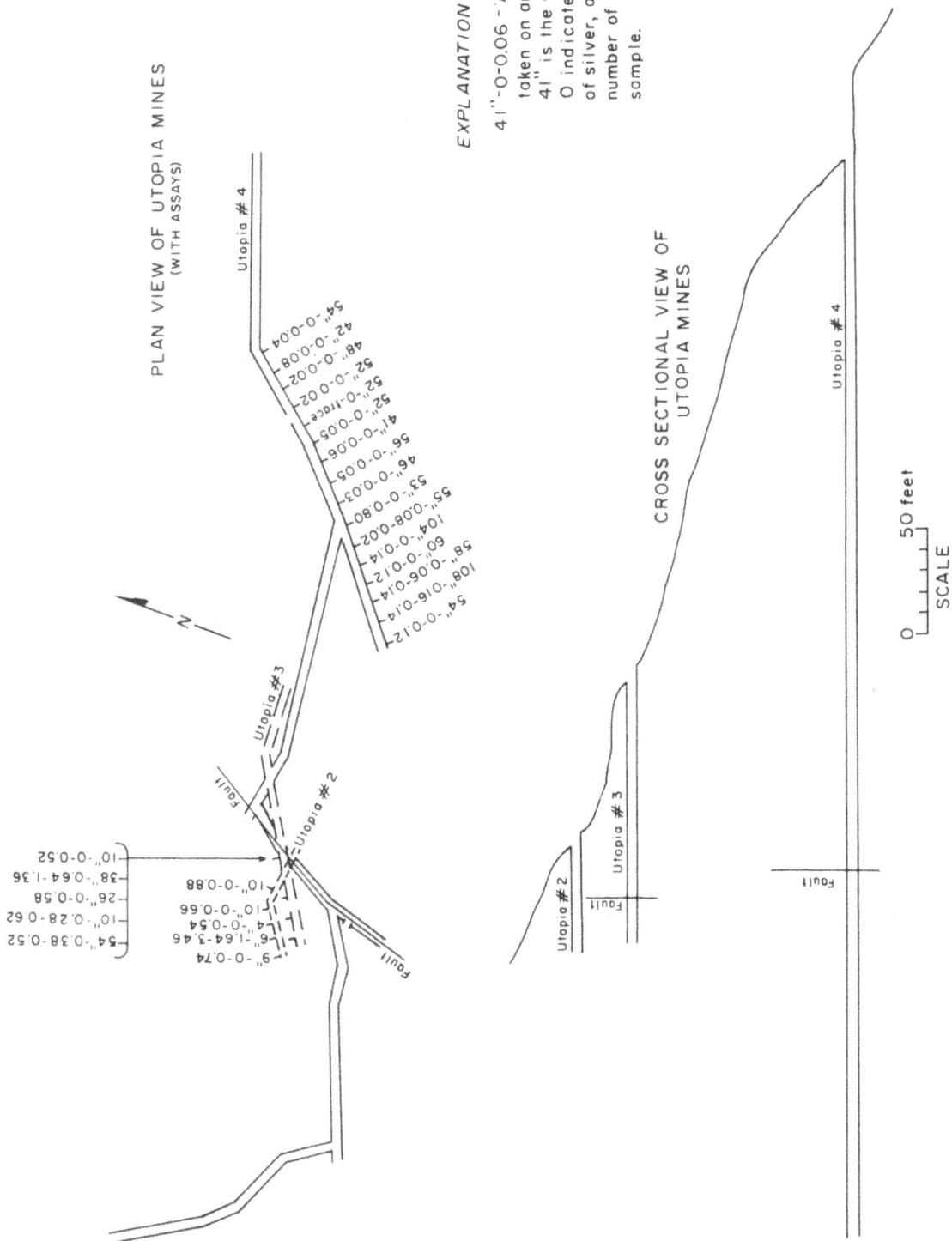

Figure 11. Utopia gold mine, Centennial Ridge district (after Dart, 1929).

Queen Mine

location: NW¼ sec. 16, T.15N., R.78W.

geology: The Queen shaft is developed in amphibolite and metagabbro
            emplaced in hornblende and felsic gneiss. The shaft exposed
            numerous small faults and veins of calcite and pegmatite.
            The faults contain up to an inch in gouge and some included
            sulfides and calcite veinlets (McCallum, 1968). A sample
            of gouge collected by Hess (1926) assayed 0.03 ounces
            platinum, 0.05 ounces iridium, less than one ounce of
            silver, and a trace of gold per ton.

Kentucky Derby Mine

location: Sec. 8, T.15N., R.78W.

geology: The Kentucky Derby Mine is developed on a few short prospect
            crosscuts. The workings penetrate strongly shattered and
            sulfide enriched dark, dense, shear-zone tectonites. The
            mineralized zone is oxidized and the sulfide masses are
            coated with films of limonite and alum. According to
            private communications to McCallum (1968), platinum group
            metals and gold were found in samples taken from this
            prospect. However, McCallum reported that grab samples
            assayed by the University of Wyoming Natural Resource
            Research Institute showed only a trace of copper, gold,
            and silver, but no platinum.

Independence Mine

location: CN½SW¼ sec. 8, T.15N., R.78W.

geology: The Independence mine is driven along a N60°E trending
            shear zone that intersects the Kentucky Derby prospects
            nearly 1/3 mile north of the workings.

            McCallum (1968) reports, "The mine was opened as a
            crosscut tunnel which extends about 95 feet into the
            shear zone and cuts five sulfide enriched veins. At the
            second vein, a distance of about 45 feet from the portal,
            a drift was driven some 25 feet to the southwest along

the mineralized trend where it deviates to the south for nearly 50 feet before it again intersects the first vein encountered in the main crosscut. The drift then parallels this vein for another 45 feet. Gold, silver, and the platinum metals are all associated with the sulfides which occur as fracture and breccia fillings, pods and small grains disseminated through tectonites."

Ores are reported to assay 0.08 ounces gold, 0.22 ounces silver, 0.27 ounces platinum, 0.46 ounces iridium, 0.03 ounces rhodium, and 0.10 ounces of osmium per ton (Hess, 1926).

Cliff Mine

location: S½SW¼ sec. 8, T.15N., R.78W.

geology: The Cliff mine was developed on a 775-foot drift which followed the schistosity of the country rock. Near the end of the main drift, a 325-foot crosscut tunnel running west-northwest intersected four quartz- and sulfide-bearing fracture zones.

The first mineralized zone was intersected a few feet from the main drift, and the quartz-sulfide vein was reported to carry $12 (0.58 ounces) of gold per ton.

The number 3 vein, located 140 feet from the main drift, was reported to carry platinum values (Hess, 1926). The wall rocks are mafic schists and submylonites (McCallum, 1968).

Empire Mine

location: NE¼NW¼ sec. 17, T.15N., R.78W.

geology: Two adits were drifted along iron and copper carbonate-stained shear planes in a shear zone parallel to the schistosity and foliation of adjacent mafic schists and gneisses (McCallum, 1968). The two drifts extended about 175 and 100 feet into the shear zone. The lower adit contained platinum-bearing sulfides.

Assays on iron-stained sulfide-bearing fractures ranged from a trace to 0.986 ounces of silver, a trace to 0.06 ounces of gold, a trace to 1.04 ounces of platinum, a trace to 63.72 (?) ounces of palladium, and a trace to 2.84 ounces of iridium per ton (Hess, 1926).

Columbine Mine

location: SW¼ sec. 17, T.15N., R.78W.

geology: The Columbine Mine was developed parallel to schistosity and cataclastic foliation in mafic schist and submylonite. Iron-stained shear planes and finely disseminated sulfides in submylonite were reported in the tunnel; however, only copper and iron values are reported from dump samples (McCallum, 1968).

Middle Fork Placers

location: W½ sec. 17 and S½ sec. 8, T.15N., R.78W. along the Middle Fork Creek.

geology: Gold and platinum have been reported from Quaternary stream sediment deposits in the Middle Fork and its tributaries. Two of the best prospects include two flat areas in sections 8 and 17 consisting of flood plain gravels and terrace deposits.

The west side of the stream in the northern flat was reported to carry considerable gold. However, recent work did not substantiate this claim (McCallum, 1968).

Queen Mill Run Placer

location: NE¼ sec. 8, T.15N., R.78W.

geology: Some gold and platinum metal were reported in stream sediments (McCallum, 1968).

Fall Creek Placer

location: SE¼ sec. 7, T.15N., R.78W.

geology: Some gold and platinum were reported in these placer deposits (McCallum, 1968).

Introduction

The Cooper Hill District, located on the northern edge of the Medicine Bow Mountains, was established in the late 1800's (Figure 12). Gold, lead, copper, and silver were produced from quartz veins with some very rich ore valued up to $84,000 per ton (1906 prices) reported (Beeler, 1906). A stamp mill was constructed near the southeastern edge of Cooper Hill in 1898 to process ore from the Albion and Emma G. mines. Several hundred tons of ore were processed, with very poor recovery results. According to

Figure 12. (a) Mine dump tailings and (b) the remains of an old stamp mill in the Cooper Hill District (photographs by W. D. Hausel, May 1980).

Beeler (1906), the tailings produced from the mill were richer than the original ores.

Beeler (1906) reported that an immense dike of "sugar" quartz on the west side of the hill consistantly averaged $2 to $2.50 (0.11 - 0.13 ounces) in gold per ton.

Location

The Cooper Hill District is located on two 1:24,000 scale U. S. Geological Survey topographical quadrangle maps: Arlington (1958) and Morgan (1961).

A geological map of Cooper Hill was prepared by Schoen (1953). The mining district lies within T.18N., R.78W., in the Medicine Bow Mountains to the northwest of Laramie (Plate 1).

Geology

The Cooper Hill district is underlain by folded Precambrian metasedimentary rocks and mafic volcanics (Figure 13). The metasediments are predominately gneisses, marbles, schists, and quartzites; the metavolcanics are amphibolites.

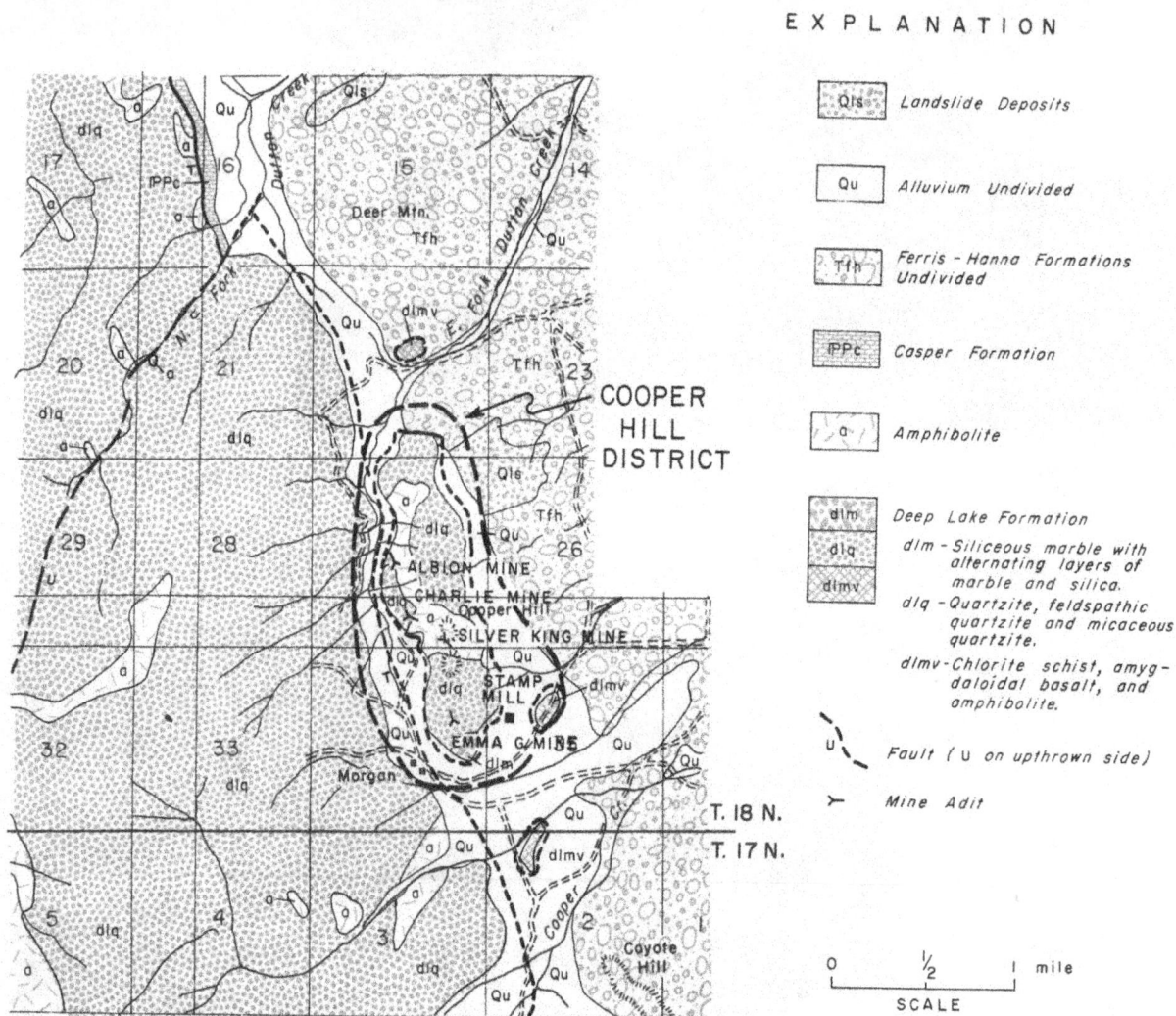

Figure 13. Geology of the Cooper Hill district (T.18N., R.78W.) (modified from Houston et al., 1968).

40

Initially during the Precambrian, sandstone, shales, and limestones were deposited in the region of Cooper Hill. These rocks were folded and faulted, producing an anticline flanked by synclines. This tectonic disturbance was followed by basalt intrusion. All these rock units were then metamorphosed to their present metasedimentary and metavolcanic equivalents.

Prior to the episode of ore mineralization, silicification resulted in the intrusion of barren quartz veins along north-south and east-west joint trends. Deformation resulted in the fracturing and displacement of many of these early formed veins. Subsequently, mineralized quartz veins intruded concordant to the foliation of the enclosing rock (Schoen, 1953).

Economic Geology

Two stages of silicification, resulting in the emplacement of quartz veins, are evident. The first stage veins are commonly displaced by faults and fractures and follow steeply dipping to vertical north-south and east-west joint trends. These veins are composed of milky quartz, and are either barren of additional mineralization or contain minor pyrite, biotite, muscovite, hornblende, and epidote gangue. In places where these veins intrude marble or calcic schist, some disseminated chalcopyrite and chalcocite commonly appear.

The second stage of mineralization follows foliation in the country rock. Where the host is quartzite, the vein may contain argentiferous galena, polybasite, pyrite, and gold as cavity fillings. In calcic schist host, ore minerals are predominately chalcocite and chalcopyrite (Schoen, 1953).

Mine Descriptions

The Emma G mine produced the richest float. The Albion mine, on the west side of Cooper Hill, produced the greatest body of gold and galena. The Richmond Mine, near the northern end of the hill, produced the largest amount of free-milling gold, and the Cooper Hill mine produced the greatest amount of copper in the district (Beeler, 1906).

Albion Mine

location: NE¼SW¼ sec. 27, T.18N., R.78W.

geology: Located on the western flank of Cooper Hill. The Albion adits follow a horizontal quartz vein at the contact between quartzite and calcic schist (Figure 14). Where the vein enters the schist, it frays out and contains some disseminated chalcopyrite, chalcocite, and bornite.

Where it lies at the contact between the two host rocks or enters quartzite, the vein contains argentiferous galena. Along most of the length of the tunnel, the vein dips slightly to the west. At two localities, approximately 70

41

LEGEND

Quartz vein

Schist

Quartzite

Strike and dip of foliation

Strike and dip of joints

Strike of vertical joints

Inaccessible workings

Shaft at level

Elev. of Portals 8641'

Well mineralized

2" quartz vein carries mineralization

Well mineralized

Vein plunges beneath adit

12" milky quartz vein carrying argentiferous galena reappears here and follows down winze parallel to foliation.

Milky Quartz veins carrying biotite, muscovite and epidote faulted by horizontal quartz vein carrying argentiferous galena.

Quartz vein following winze widens to 2' but shows no mineralization.

End of drift

East
Back of stope
West

End of adit

0   10   20   30 feet

Scale

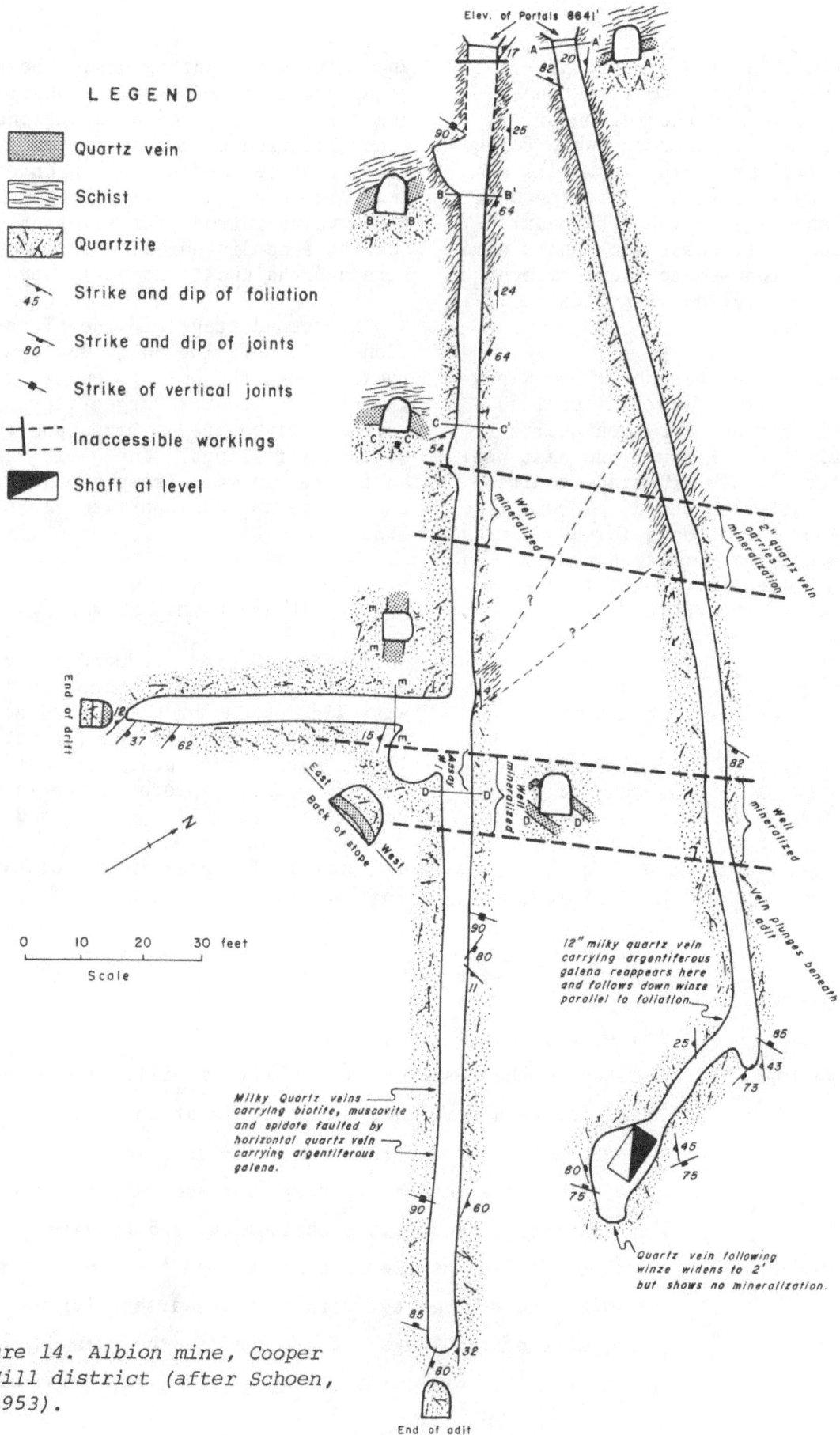

Figure 14. Albion mine, Cooper Hill district (after Schoen, 1953).

42

feet and 100 feet from the portal, the vein rolls (changes
dip to the east), forming a favorable condition for ore
deposition. An assay at the second vein roll gave 0.825
percent lead, 0.70 ounces of gold, and 2.20 ounces of silver
per ton (Schoen, 1953). Osterwald et al. (1966) reported
selected assays from a 5-to-9-foot-wide, cerussite-bearing
quartz vein to contain from 4 to 5.3 ounces of gold and
50 ounces of silver per ton, in addition to lead.

Charlie Mine

location: $S\frac{1}{2}S\frac{1}{2}$ sec. 27, T.18N., R.78W.

geology: The Charlie mine is a short adit developed on a horizontal
vein containing specular hematite. No ore minerals are
reported (Schoen, 1953).

Silver King Mine

location: $S\frac{1}{2}NE\frac{1}{4}$ sec. 27, T.18N., R.78W.

geology: A short adit is developed in biotite schist. No evidence
of mineralization is reported on the property.

Emma G. Mine

location: $N\frac{1}{2}SE\frac{1}{4}$ sec. 34, T.18N., R.78W.

geology: Beeler (1906) reported this property to contain some of
the richest ore in the district. The property is developed
by a vertical shaft and a short inclined shaft. The wall
rock is pyroblastic biotite schist.

It was reported that near the bottom of the inclined
shaft, a quartz vein reaches more than 15 feet thick --
however, it is unmineralized (Schoen, 1953).

Fox Group

geology: A 70 foot shaft was sunk on a 12-foot sulfide "body".
Black sulfides and a thin layer of tellurium were discovered
at the shaft's bottom. Reported assays gave 1.15 to 2.5
ounces of gold per ton (Osterwald et al., 1966).

GOLD HILL DISTRICT

The Gold Hill District was established in 1890 near Medicine Bow Peak in the Snowy Range. The district lies nearly 45 air miles west of Laramie in T.16N., R.80W. (Plate 1).

During 1905, plans to extensively develop the Gold Hill property were under consideration. The plans included an extension of a railroad spurr from Centennial to Gold Hill, and the construction of a townsite of twenty-five cottages, a hotel, and a power plant for generation of electricity (The Centennial Post, Dec. 30, 1905). Expensive boilers and steam operated hoists were installed at some of the mine sites (Childers, 1957).

The total gold production from the district is not known. However, at least $3000 (145 ounces) in gold had been produced by 1893 (Knight, 1893). The Gold Hill ore was reported to be rich and to occur in quartz veins primarily in norite country rock. The most extensive mine development was at the Acme Shaft (sec. 15, T18N., R.80W.) (Figure 15) which reached a total depth of 169 feet in 1905 and was reported to be in rich ore at that time (The Centennial Post, Dec. 30, 1905).

JELM MOUNTAIN DISTRICT

Introduction

In 1872, the Bramel Mining District (better known as the Jelm District) was organized to include Jelm Mountain and the Boswell Creek area to the west. Several prospects and mines were developed on auriferous quartz veins and shears, and three gold stamp mills were constructed near the present site of Woods Landing (Michalek, 1952). As the gold veins began to give out, the district began producing an increasing amount of copper. The early prospects were developed on oxidized copper-gold veins. Near the surface, much of the copper was removed by downward-leaching supergene solutions, leaving behind oxidized gold-quartz veins. As mining reached to shallow depths, the supergene copper deposits were developed, and copper became the more important ore. Assays as high as 30 percent copper were reported, but apparently the supergene enrichment was restricted to small tonnages, and the ore played out. In a short time the district was abandoned (Figure 16).

Location and Accessibility

The Jelm Mountain District lies 30 miles west-southwest of Laramie by Highway 230 (Plate 1). From the junction of 230 with Highway 10 at Woods Landing, the district is entered by driving three miles south on Highway 10 to where Highway 10 is intersected by the observatory road (an unimproved dirt road) heading east (Figure 17).

The district lies within the boundaries of the Jelm Mountain (1963) U. S. Geological Survey topographic quadrangle. Geological maps of the district were prepared by Michalek (1952) and King (1961).

Geology

Jelm Mountain is an uplifted block fault bounded on the western and eastern flanks by westward dipping thrust faults. The thrust on the east flank dips 60 degrees west and may have more than 2,000 feet of displacement --

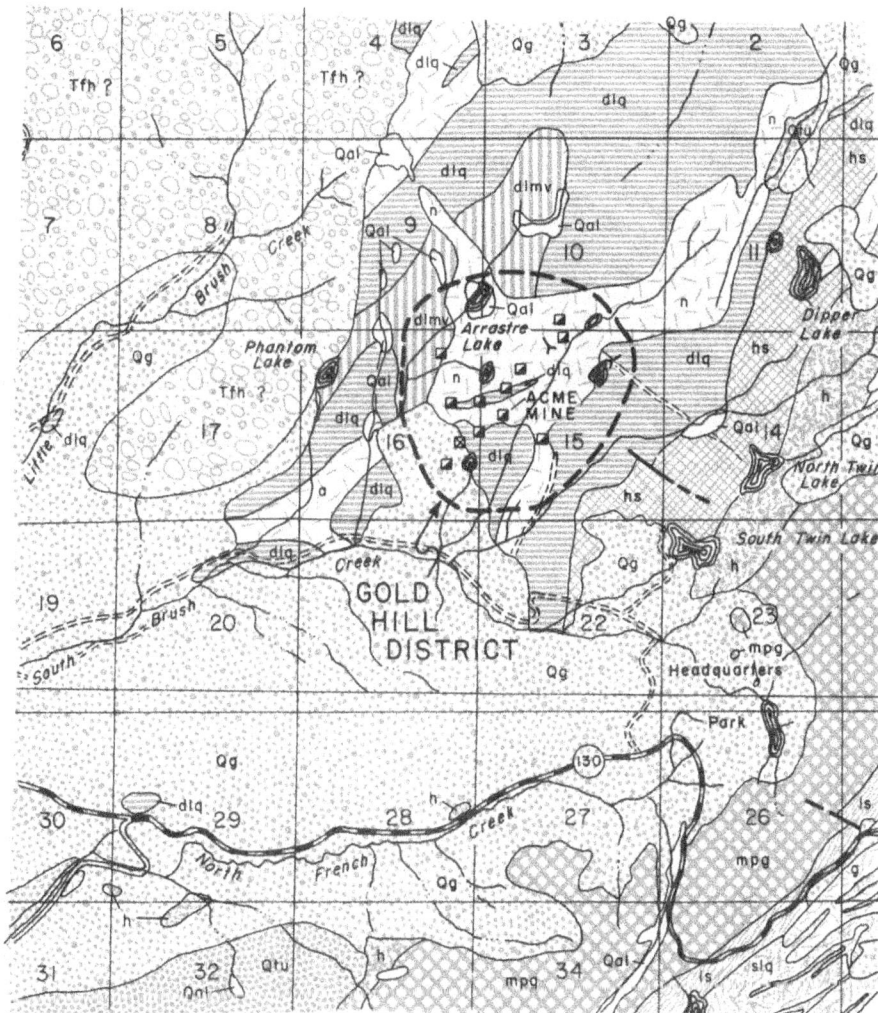

Figure 15. General geology of the Gold Hill district (T.16N., R.80W.) (modified
from Houston et al., 1968).

Figure 16. Mine dump tailings in the Jelm Mountain District (photograph by W. D. Hausel, May 1980).

Precambrian rocks have been thrust over the top of Paleozoic sediments. Near the southeastern edge of the district, exposed mine dump wastes of Precambrian gneiss and schists along with fragments of the Pennsylvanian Fountain Formation suggest that the shaft penetrated the thrust and bottomed out in the Paleozoic section (Michalek, 1952).

The predominate rocks of Jelm Mountain are schists, gneisses, pegmatites, and granitic rocks; these rocks are completely surrounded by Paleozoic, Mesozoic, Tertiary, and Quaternary sediments and alluvium. The schistosity of the Precambrian metasediments strikes predominately east-west.

Economic Geology

Copper and gold prospects occur in quartz veins, shear zones, pegmatite, and gneissec bodies (Osterwald et al., 1966). Many of the prospects are stained by hematite, limonite, malachite and azurite. Identified ore minerals include cuprite. chalcopyrite, pyrite, native gold, native copper, pyrite, and galena. Michalek (1952) proposed that the gold originated from auriferous soda-rich granite, and the copper from basic intrusives.

Mine Descriptions

Annie Mine

geology: The Annie Mining Company was reported by Beeler (1906) to have developed a 3- to 4-foot wide quartz lead by a 140-foot-deep shaft intersecting a 138-foot adit at 60-feet

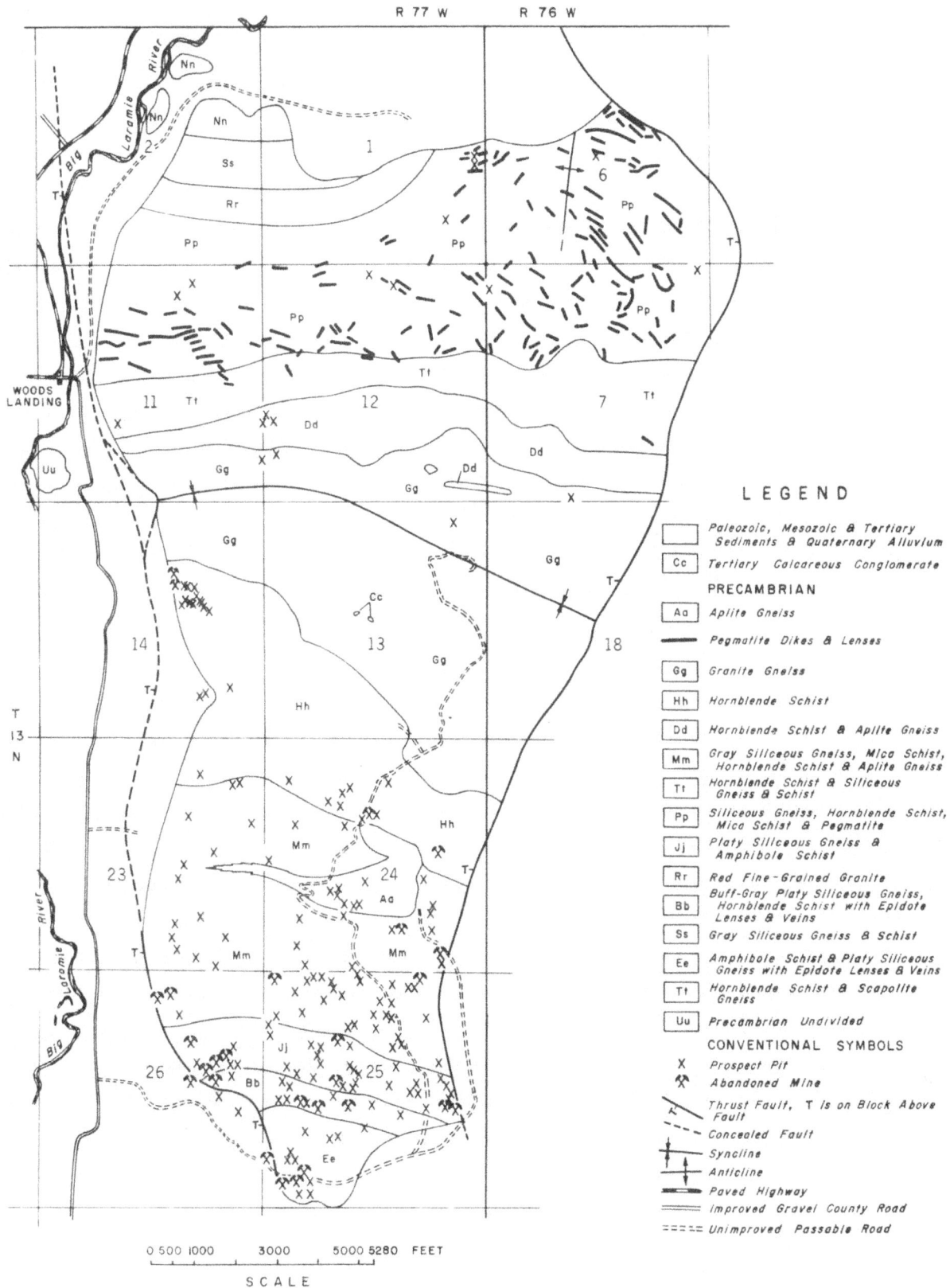

LEGEND

| | Paleozoic, Mesozoic & Tertiary Sediments & Quaternary Alluvium |
| Cc | Tertiary Calcareous Conglomerate |

PRECAMBRIAN

Aa	Aplite Gneiss
—	Pegmatite Dikes & Lenses
Gg	Granite Gneiss
Hh	Hornblende Schist
Dd	Hornblende Schist & Aplite Gneiss
Mm	Gray Siliceous Gneiss, Mica Schist, Hornblende Schist & Aplite Gneiss
Tt	Hornblende Schist & Siliceous Gneiss & Schist
Pp	Siliceous Gneiss, Hornblende Schist, Mica Schist & Pegmatite
Jj	Platy Siliceous Gneiss & Amphibole Schist
Rr	Red Fine-Grained Granite
Bb	Buff-Gray Platy Siliceous Gneiss, Hornblende Schist with Epidote Lenses & Veins
Ss	Gray Siliceous Gneiss & Schist
Ee	Amphibole Schist & Platy Siliceous Gneiss with Epidote Lenses & Veins
Tt	Hornblende Schist & Scapolite Gneiss
Uu	Precambrian Undivided

CONVENTIONAL SYMBOLS

X	Prospect Pit
⚒	Abandoned Mine
	Thrust Fault, T is on Block Above Fault
---	Concealed Fault
	Syncline
	Anticline
	Paved Highway
	Improved Gravel County Road
=====	Unimproved Passable Road

0 500 1000   3000   5000 5280 FEET

SCALE

Figure 17. Jelm Mountain District (after Michalek, 1950).

47

depth. The wall rock in the mine is granite and diorite associated with hornblende and tourmaline schists.

Ore mineralization included copper sulfides and gold. Assays were reported to range between 3 to 30 percent copper and 0.09 to 0.15 ounces of gold per ton (Beeler, 1906).

The Wyoming Queen Mining Company

geology: Three shafts were developed, known as the Colorado, Boston, and Rising Sun mines. The Colorado shaft, located about 3,000 feet north of the Boston shaft, was developed to 250 feet deep on a vein with copper and iron sulfides and stringers of native copper. The Boston shaft reached at least 80 feet deep and was developed on a gold vein. The Rising Sun was developed on a vein parallel to the Boston and Colorado veins, containing galena with values in gold and silver (Beeler, 1906).

ABSAROKA MOUNTAINS

The Absaroka Mountains are located in extreme northwestern Wyoming, on the eastern boundary of Yellowstone Park. This range is, in general, a deeply dissected plateau formed by the accumulation of several thousands of feet of layered calc-alkalic flows, breccias, and tuffs of basaltic to rhyolitic composition. These extrusive volcanics have been intruded by stocks, plugs and dikes ranging in composition from syenite and granodiorite to gabbro.

During the Tertiary, magmas poured out and were ejected from several volcanic centers within the range. These centers are intensively altered, and contain disseminated mineralization and networks of fractures filled with stockwork veinlets. Such an association is typical of porphyry copper-molybdenum deposits which supply a tremendous amount of base and precious metals to the world market. Mineralized sites in the Absarokas include the Kirwin District (Wood River), the Meadow Creek area, the Sunlight Basin District, the Stinkingwater District, the Silver Creek area, and the Yellow Ridge area (Plate 1). Undoubtedly, additional porphyry sites will be located in the Absaroka Mountains in the future.

THE SUNLIGHT REGION

Introduction

The Sunlight Mining Region lies about 50 miles northwest of Cody, Wyoming, within the Absaroka Range (Figure 18). The region is rugged; however,

48

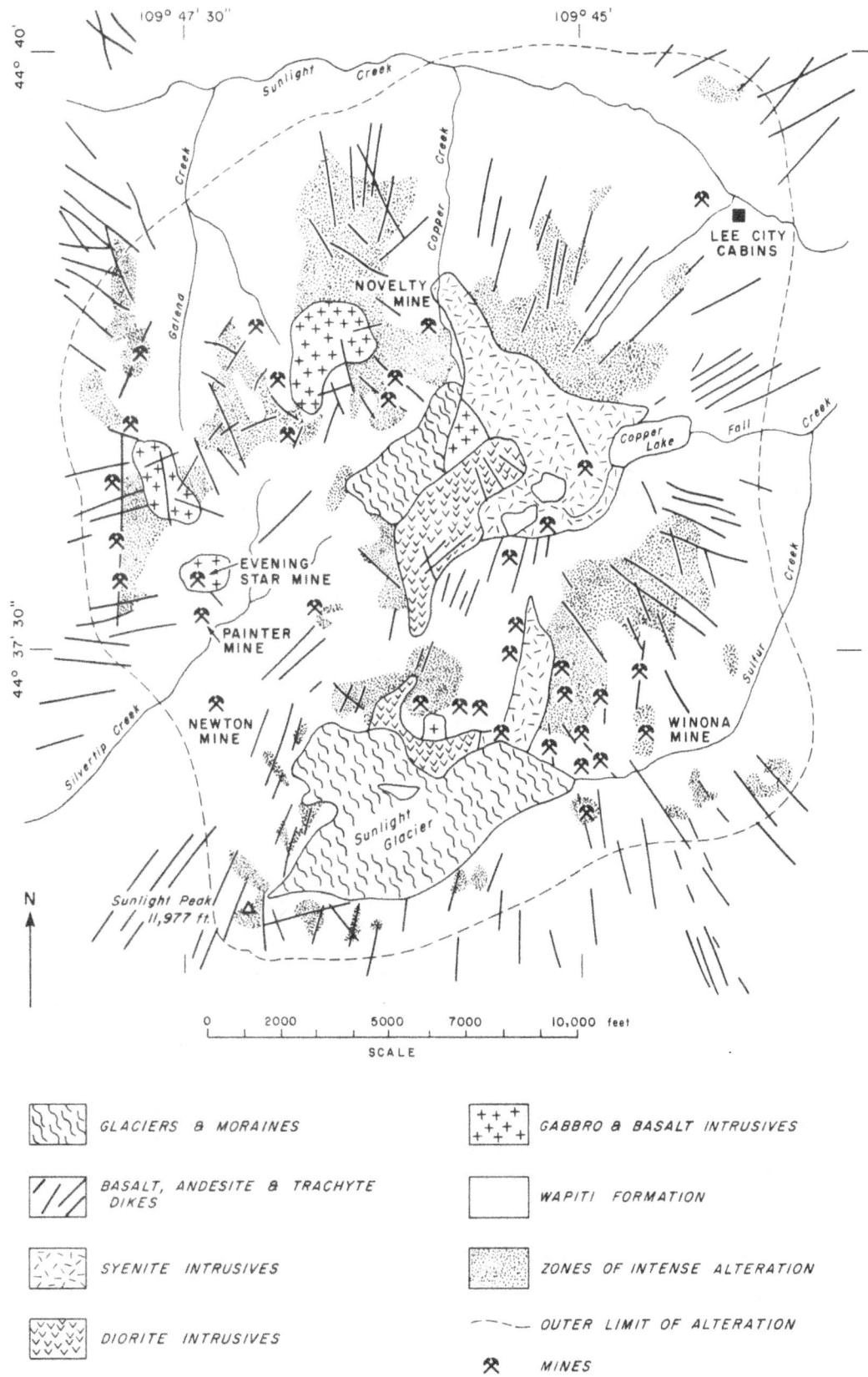

Figure 18. General geology of the Sunlight District, Park County, Wyoming (after Parsons, 1937).

one can drive within ten miles of the district on good paved roads. The last few miles are on a rugged dirt road that crosses several gravel-bottom streams, and on steep pack trails. Osterwald et al. (1966) and Parsons (1937) gave excellent descriptions of the mineralization in this region. Rich (1974), Petersen (1968), and Dreier (1967) conducted thesis problems on portions of the district.

Geology

The Tertiary volcanics of the Sunlight district unconformably overlie Paleozoic rocks. Tertiary volcanics and shallow intrusives are the only rocks exposed in the district. The volcanics are composed of extrusive pyroclastics and interbedded flows which have been named the Wapiti Formation (Rich, 1974).

Economic Geology

Mineralization in the Sunlight Mining region occurs as vein-filled fissures and as disseminated replacements in andesitic volcanics near both syenite and monzonite intrusive stocks.

Veins are subdivided into two separate groups -- those that are essentially barren and those that are mineralized. The barren veins occur as quartz-pyrite, pyrite, magnetite, and carbonate veins, with some minor chalcopyrite commonly associated with the pyrite- and magnetite-bearing veins.

Mineralized veins occur primarily

as either copper-bearing or lead-silver-bearing veins. The copper-bearing veins contain chalcopyrite, pyrite, bornite, tetrahedrite, native gold, sphalerite, and galena as the primary ore minerals. Minerals in the supergene zone include limonite, malachite, covellite, and chalcocite. The gangue minerals are commonly quartz, limonite, adularia, and carbonates.

The lead-and-silver-bearing veins commonly contain galena, tetrahedrite, minor stromeyerite, chalcopyrite, and native silver, with calcite, ankerite, siderite, and barite gangue (Elliott, 1980).

The disseminated deposits consists of (1) pyrite in altered zones, (2) magnetite and chalcopyrite in intrusive rocks, and (3) stockworks of copper-bearing veins and veinlets (Elliott, 1980). Both (1) and (2) disseminated deposits occur primarily as replacement mineralization, whereas the stockworks (3) occur both in fissure fillings and as replacements.

Parsons (1937), Dreier (1967), Petersen (1968), Rich (1974) and Elliott (1980) all recognized the intensive alteration associated with the mineral deposits of the region. Similar alteration is found associated with most of the world's copper-molybdenum porphyry deposits. Near the intrusives and adjacent to veins and veinlets, intense propylitic and potassic alteration is dominant. Propylitic alteration is more extensive than the potassic alteration, which is restricted to the area immediately adjacent ot the intrusives. Argillic alteration is found in the outer regions of the intensely altered zones.

Mine Descriptions

Copper Lake Deposits

geology: These deposits are fissure filling veins in volcanic breccia near the stock contacts. Common minerals are pyrite, wolframite, tetrahedrite, and lesser sphalerite. The gangue is quartzose. Dump samples average 0.54 ounces of gold and 13.46 ounces of silver per ton (Parsons, 1937).

In the same general region that Parsons (1937) reported assays, Williams (1980) reported assays from several prospect pits and dumps to range from 0.005 to 0.06 ounces of gold and a trace to 24.32 ounces of silver per ton, a trace of 0.08 percent copper, 0.028 to 0.10 percent lead, and 0.02 to 0.30 percent zinc. The analyses did not include tungsten.

Copper Creek Adit

geology: The Copper Creek adit was driven into the side of a cliff in the Copper Creek valley about 1½ miles due north of Stinkingwater Peak. The adit was driven nearly 200 feet through andesite to intersect a sheared mineralized syenite dike. The dike was sampled and assayed a trace to 0.01 ounces of gold and a trace to 0.10 ounces of silver per ton, a trace to 0.01 percent copper, 0.02 to 0.04 percent lead, and 0.04 to 0.05 percent zinc.

Additional samples taken from nearby prospects assayed from 0.005 to 0.01 ounces of gold and a trace to 0.10 ounces of silver per ton, a trace to 0.02 percent copper, 0.035 to 0.04 percent lead, and 0.04 to 0.10 percent zinc (Williams, 1980).

Galena Creek Basin Prospects

geology: Galena Creek Basin contains several prospects and adits developed at the contacts between syenite dikes and the andesitic country rock.

Reported assays from these prospects range from a trace to 0.01 ounces of gold and 0.29 to 0.80 ounces of silver per ton, a trace to 0.01 percent copper, 0.02 to 0.26 percent lead, and 0.05 to 0.12 percent zinc (Williams, 1980).

Hoodoo Claim

geology: Calcite and pyrite occur as secondary wall rock minerals
and contain coarse galena in iron carbonate gangue. An
assay yielded 0.02 ounces of gold and 1.98 ounces of silver per ton, and 9.4 percent lead with a trace of copper
(Parsons, 1937).

Winona Claim Group

The Winona claim group consists of 31 claims located between 1901 to 1910
in the Sulfur Creek valley. Several of these claims were patented (Williams,
1980; Rich, 1974). The Winona claim group included the following claims:
Doubtful, Gopher, Hidden Treasure 1, Malachite 1, Hidden Treasure, B and S,
Malachite, Butte, Greenhorn, and Copper King. Rich (1974) reported assays as
follows:

Claim	Copper percent	Gold ounces	Silver ounces
Doubtful	29.2	0.28	19.00
Gopher	6.7	0.24	1.76
Hidden Treas. 1	47.8	0.16	11.64
Malachite 1	44.7	0.12	25.46
Hidden Treas.	12.4	0.05	-----
B & S	25.8	0.08	16.40
Malachite	34.8	0.08	30.50

Winona Mine (Greenhorn Claim)

geology: An assay of dump material from this mine was reported to
average 0.58 ounces of gold and 0.88 ounces of silver per
ton (Parsons, 1937).

Butte Claim

geology: Assays on this vein yielded 0.22 ounces of gold and 3.14
ounces of silver per ton, and 4.43 percent copper (Parsons,
1937).

Copper King Claim

geology: These veins range up to 10 feet in thickness and occur as fissure fillings in shear zones and faults in volcanic breccia. Gangue minerals are quartz and carbonate. Ore minerals are tetrahedrite, famatinite, bornite, and galena. Dump sample assays averaged 0.04 ounces of gold and 0.62 ounces of silver per ton, and 3.35 percent copper (Parsons, 1937).

Big Goose Vein

geology: This is a fissure-fill vein in a shear zone at the contact between volcanic breccia and a dike. Ore minerals include chalcopyrite, pyrite, and traces of tetrahedrite, galena, and sphalerite. Sylvanite and rare grains of native gold are scattered throughout the vein. An assay gave 0.82 ounces of gold per ton and a trace of silver (Parsons, 1937).

Tip Top Claim

geology: Gangue minerals from this claim include calcite, ankerite, siderite, and barite. Ore minerals include galena and argentiferous tetrahedrite associated with proustite and stromeyerite. Small amounts of chalcopyrite and pyrite are present, and the wall rocks are sericitized. An assay of dump material yielded 0.02 ounces of gold and 7.14 ounces of silver per ton, 0.60 percent lead, and a trace of copper (Parsons, 1937).

Morning Star Claim

location: Located approximately 1500 feet north of the Evening Star Mine at an elevation of 10,650 feet on the U. S. Geological Survey Sunlight Basin (1:62,500) topographic quadrangle.

geology: This claim is developed on a sheared syenite dike carrying ankerite, siderite, calcite and quartz gangue. Some scattered argentiferous galena contains 0.23 ounces of silver and 0.01 ounces of gold per ton (Parsons, 1937).

Williams (1980) estimates that 825 tons of ore were mined from this prospect.

Evening Star Claim

location: The Evening Star Claim is located on an arete above Silvertip Basin at nearly 10,570 feet elevation. The Evening Star Mine is found on the Sunlight Peak topographic quadrangle.

geology: The Evening Star Mine was developed prior to 1901 (Rich, 1974). Assays from the outcrop vein gave 0.02 ounces of gold and 91.44 ounces of silver per ton, 1.40 percent copper, and 1.30 percent lead (Parsons, 1937).

Two grab samples taken from the mine dump, reported by Williams (1980), yielded 0.005 and 0.01 ounces of gold per ton, 3.19 and 4.20 ounces of silver per ton, 0.02 and 0.025 percent copper, 0.04 and 0.06 percent lead, and 0.03 and 0.05 percent zinc.

Parsons' sample was probably a selected sample which would explain the high silver content.

Hardee's Claim

geology: An assay of dump material yielded 0.04 ounces of gold and 20.24 ounces of silver per ton, 62.23 percent lead, and a trace of copper (Parsons, 1937).

Newton Prospect

location: Located at 9,266 feet elevation in Silvertip Basin.

geology: The mine was developed in andesite porphyry to more than 200 feet into the mountain side with several shorter drifts extending from the main adit. Ore minerals include chalcopyrite, chalcocite, and malachite (Rich, 1974).

Williams (1980) assayed a grab sample from the prospect and reported it to yield 0.005 ounces in gold and 0.49 ounces in silver per ton, 0.03 percent copper, 0.04 percent lead, and 0.02 percent zinc.

MeClung Mine

    location: At the base of Stinkingwater Peak.

    geology: A 150-foot tunnel showed galena and chalcopyrite in a fissure vein cutting quartz porphyry. The vein carries some gold and silver (Osterwald et al., 1966).

Novelty Mine

    location: Located at the headwaters of Galena Creek near the base of the Galena Creek glacier at 9,204 feet elevation.

    geology: The mine was developed on a syenite dike in andesite country rock (Williams, 1980).

           Assays reported by Parsons (1937) of the dump material yielded 0.06 ounces of gold and 19.0 ounces of silver per ton, 13.10 percent lead, and 0.75 percent copper.

           Williams (1980) reported assays to range from a trace to 0.02 ounces of gold per ton, a trace to 1.50 ounces of silver per ton, a trace to 0.03 percent copper, 0.02 to 0.88 percent lead, and 0.05 to 0.20 percent zinc.

Painter Mine (Silvertip Group)

    location: Located on the northwestern side of Silvertip Basin at 9,370 feet elevation.

    geology: This property was claimed in 1890 and patented in 1907. The last ore shipment was reported in 1903, when 100 tons of ore were sold by the Sunlight Mining Company (Williams, 1980). Assays of dump material at the mine yielded an average of 0.06 ounces of gold and 5.44 ounces of silver per ton, 3.85 to 14.9 percent copper, and 10.6 percent lead. It is estimated that the Painter deposit contains $7,300,000 in ore reserves (Rich, 1974).

           Assays reported by Williams (1980) were significantly lower than those reported by Rich. Gold was reported to range from 0.005 to 0.06 ounces per ton, and silver ranged between 0.47 and 2.45 ounces per ton, with 0.12 to 2.30 percent copper, 0.02 to 0.08 percent lead, and 0.04 to 0.10 percent zinc.

Upper Silvertip Basin Prospect

location: The adit is located at an elevation of nearly 10,000 feet on the east side of the basin.

geology: The property was developed by a N85°W-bearing adit extending 52 feet into blocky andesite. A sample of altered material from the mine dump assayed 0.005 ounces in gold and 0.30 ounces in silver per ton, 0.03 percent copper, 0.04 percent lead, and 0.10 percent zinc (Williams, 1980).

Lee City Prospect

location: At Lee City on Sunlight Creek.

geology: An adit was developed bearing N20°E. A grab sample of dump material assayed 0.005 ounces of gold and 0.20 ounces of silver per ton, 0.01 percent copper, 0.02 percent lead, and 0.02 percent zinc (Williams, 1980).

STINKINGWATER DISTRICT

The Stinkingwater District is located in sec. 18, T.47N., R.106W. The district contains multiple intrusions of granodiorite to diorite which have domed, fractured, and metamorphosed igneous rocks of the Wapiti and the Wiggins formations. The Stinkingwater mineralized area covers approximately 4 square miles on the northwest slope of Crater Mountain (Osterwald et al., 1966).

Mineralization in the district is principally disseminated copper-molybdenum. Outcrops are highly fractured and bleached from the emplacement of stockworks and veins. Some supergene enrichment is present on the northwest side of Crater Mountain (Fischer, 1972).

KIRWIN DISTRICT AND UPPER WOOD RIVER AREA

The Kirwin District and Upper Wood River area have mineralization similar to that of the Sunlight Basin and Stinkingwater districts. Wilson (1960; 1964; 1975) describes the geology, mineralogy, and alteration of the region in detail. The country is very rugged, with nearly 3,000 feet of relief. The elevation of Wood River is about 9,200 feet and adjacent ridges reach to more than 12,000 feet.

Rock units are primarily of the Wiggins Formation intruded by granodiorite stocks, andesite plugs, and numerous andesite and dacite dikes. Osterwald et al. (1966) describe the veins as fissure fillings containing quartz gangue, pyrite, galena, sphalerite, chalcopyrite, chalcocite, tetrahedrite, molybdenite, stephanite, and specular hematite. The oxidized zone contains limonite, malachite, anglesite, molybdenite, native copper, azurite, cuprite and native gold.

The U. S. Bureau of Mines (1979) reported the Kirwin prospect to contain 63,500,000 tons of demonstrated (measured and indicated) copper reserves at an average grade of 0.75 weight percent. In addition, 63,500,000 tons of 0.015 percent molybdenum were reported. Although no estimates of gold and silver are included, this deposit would also contain a fair reserve of these precious metals.

SILVER CREEK AND YELLOW RIDGE AREA

The Silver Creek and Yellow Ridge areas were discovered in the 1970's by the U. S. Geological Survey during a mineral-evaluation study in the Washakie Wilderness.

The Silver Creek area is located on the Fall Creek 1:24,000 scale topographic quadrangle and the Yellow Ridge area is found on the Francs Peak 1:24,000 quadrangle.

In mineralization and alteration, these two regions are similar to the Sunlight Basin, Stinkingwater region, and Kirwin and Meadow Creek areas, where typical copper porphyry mineralization and associated stockworks have been identified. In addition to copper, molybdenite, pyrite, zinc, lead, silver, and gold have been identified in anomalous amounts (Fisher et al., 1977).

GOLD-BEARING CONGLOMERATES OF NORTHWESTERN WYOMING

Introduction

Several sedimentary formations in northwestern Wyoming contain significant resources of gold-bearing quartzite conglomerates and sandstone (Antweiler and Love, 1967; Lindsey, 1972; Love, 1973; Antweiler et al., 1977). Associated alluvial deposits derived from these conglomerates and sandstones are commonly enriched in gold. The conglomerates and sandstones were transported by high-velocity eastward flowing streams originating in a now-vanished mountain range called the Targhee uplift (Love, 1973). This prehistoric uplift was located in eastern Idaho and west of Wyoming's Teton Range.

Placer gold deposits are found in terrace and modern alluvial stream deposits along the Snake, Hoback, Gros Ventre, northern Green, and Wind rivers. Characteristics of gold found in alluvial deposits in the Snake River as far west as Boise, Idaho, and in the Green River as far south as Green River, Utah suggest that this gold was derived from northwestern Wyoming gold-bearing sediments (Antweiler et al., 1977).

Total gold production from these sediments is not known. The only known production estimate for the region was estimated for the Snake River placers. These placers were reported to have produced more than 100,000 ounces of gold (Staley, 1946). Although Staley's estimate was primarily for Idaho, Antweiler et al. (1977) suggested that a substantial amount of the Snake River production was from the Wyoming placers.

The total gold resource in the northwestern Wyoming region is significant. Maguire (in Antweiler and Love,

1967) estimated the total gold resource in the Snake River gravels of Wyoming to be at least 100 million ounces. Additionally, the total gold resource of the Pass Peak Formation, which is only one of several gold-bearing sedimentary formations in the region, is estimated at more than 46 million ounces (Antweiler et al., 1977).

Location and Accessibility

The gold-bearing gravels and conglomerates cover an extensive region in northwestern Wyoming (Plate 1). These deposits extend south from the southern portion of Yellowstone Park to the Hoback River in Sublette County. A large portion of the gold-bearing conglomerates lies within the Teton wilderness and Yellowstone Park, and is inaccessible to mining.

Geology

Northwestern Wyoming contains several gold-bearing quartzite conglomerates in latest Cretaceous and Tertiary formations, including Quaternary gravels.

The late Cretaceous Harebell Formation has a maximum thickness of about 11,000 feet (Love, 1973). Gold is most abundant in the sandstone units. In the conglomeratic units, gold is more abundant in ferruginous varieties having some round quartzite stones larger than 3 inches in diameter (Antweiler and Love, 1967).

The Cretaceous Pinyon Conglomerate has a maximum thickness of about 4,000 feet (Love, 1973). Gold, again, is more abundant in the sandstone units of this formation, with the highest values reported from carbonaceous sandstone (Antweiler and Love, 1967).

The Fort Union Formation of Paleocene age in the Bighorn Basin is an approximate equivalent of the Pinyon Conglomerate. Gold content in the Fort Union is much less than in the Pinyon Conglomerate.

Several hundred feet of quartzite conglomerate of (?)early Eocene age overlies the Fort Union Formation in the Bighorn Basin. One gold analysis of this conglomerate was higher than those of any of the older formations located in the Bighorn Basin and to the west of this unit.

The Eocene Wind River Formation along the Continental Divide between Jackson Hole and the Wind River Basin contains an average of twice as much gold per ton of any other formation included in this discussion. These rocks include several quartzite conglomerates and brown sandstones similar to the Pinyon Conglomerate.

The Eocene Pass Peak Formation located along the south margin of the Gros Ventre Mountains contains more than 2,000 feet of conglomerate. These conglomerates are believed to have been reworked from conglomerates in the Pinyon and Harebell Formations. The gold content of the Pass Peak conglomerates is less than that of any of the other formations except the Fort Union.

A Miocene (?) conglomerate overlying the Pinyon Conglomerate contains slightly less gold per ton than the Pinyon. This conglomerate is about 100 feet thick and contains rounded quartzite boulders of two to three feet in diameter.

The Quaternary alluvial deposits offer the best source of placer gold in the region. These deposits were formed by reworking the conglomerates of the gold-bearing formations and depositing the gold-bearing gravels and

and sands in modern stream channels (Antweiler and Love, 1967).

Gold Content and Form

Much of the gold from the gold-bearing conglomerates is very fine material and has a tendency to float on the surface of water when agitated. This makes it difficult to recover much gold by the typical placer mining methods (e.g. panning, sluicing, Humfrey cyclone, Wilfley table) (Antweiler & Love, 1967). Table 3 gives the average gold content of the conglomerates in northwestern Wyoming.

Table 3. Reported gold content of the gold-bearing conglomerates, northwestern Wyoming (after Antweiler and Love, 1967).

Formation	Average gold content in ppb	Single-sample maximum in ppb
Quaternary Alluvium	103	2000
Miocene (?) Conglomerate	65	290
Pass Peak Formation	47	250
Wind River Formation	222	2000
Early (?) Eocene Conglomerate	94	400
Pinyon Conglomerate	86	6000
Fort Union Formation	35	300
Harebell Formation	65	1000

Mine Descriptions

Davis Claim

location: SW¼ T.38N., R116W.

geology: This is a placer prospect located on the east side of the
              Snake River about ½ mile north of Bailey Creek. Terrace
              and channel fill gravels contain flour-gold that is diffi-
              cult to recover. A terrace at the mouth of Pine Creek,
              located about 40 to 50 feet above water level and extending
              down the creek for nearly one mile, contains from a trace

to 0.11 ounces in gold per cubic yard of gravel (Osterwald et al., 1966).

Horse Creek

location: T.34N., R.114-115W.

geology: A sample from a prospect pit in Jurassic sediments assayed $66.00 in gold and $18.00 in silver per ton (1907 prices) (Schultz, 1907).

Pine Bar Diggings

location: NW¼ T.37N., R116W.

geology: At the mouth of Pine Creek on the south side of the Snake River. Fine flake gold is found in gravels below eight feet of barren material (Osterwald et al., 1966).

SEMINOE DISTRICT

Introduction

In 1872, a group of prospectors led by General L. P. Bradley and Captain T. B. Deweese set out in discovery of reported rich silver veins in the Seminoe Mountains. Instead of silver, they found several deposits containing gold and copper. Some of the veins contained free-milling gold. Development commenced, and three stamp mills were erected to recover gold. However, only about $10,000 (530 ounces) in gold were recovered (Knight, 1893). After 1900, the district was abandoned until the Sunday Morning Mine was reopened for a short interval of time in 1920 (Bishop, 1964).

Location and Accessibility

The Seminoe District is located in T.26N., R.85W. and is entered by paved and graded dirt roads heading north out of Sinclair towards the Seminoe Dam (Plate 1). U. S. Geological Survey topographic quadrangles Seminoe Dam (1953) and Bradley Peak (1953) include the Seminoe District.

Bishop (1964) mapped the district, differentiating the Precambrian rocks and showing locations of mines and of asbestos and jade deposits.

Economic Geology

The Seminoe district is divided into metasediments, metamorphic mafic rocks, ortho-amphibolite, meta-diabase, a granite intrusive, pegmatite, quartz veins, and diabase (Figure 19).

The granite is intrusive into the metasediments, which are moderately folded and faulted and lie on the northern and southwestern flanks of the district. The gold mineralization occurred as fissure fillings in quartz veins. Copper deposits also occur in quartz veins and disseminated in altered amphibole schist (Bishop, 1964).

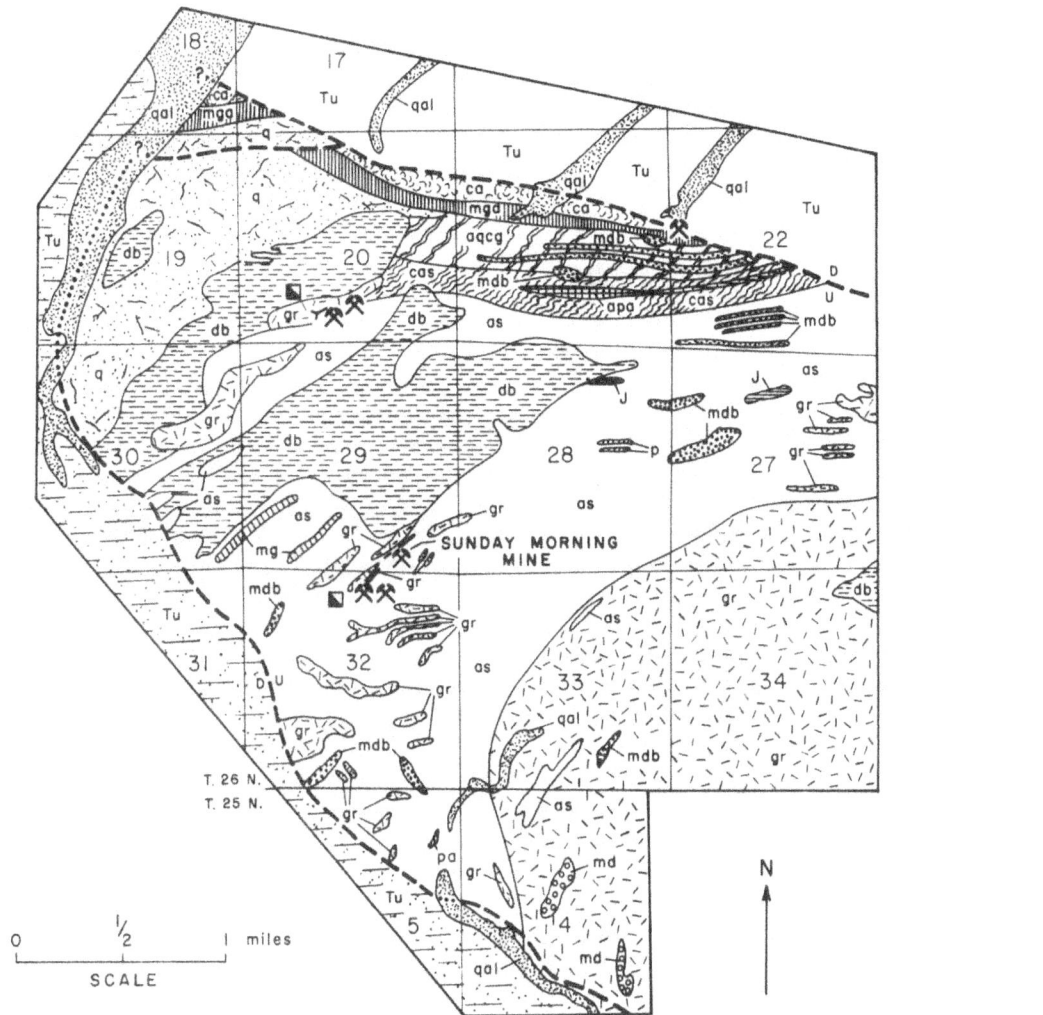

EXPLANATION

qal Quaternary alluvium	gr Granite	mga Meta-gabbro amphibolite	as Amphibole schist
Tu Tertiary undivided (includes the Moonstone and Split Rock formations)	md Meta-diorite	pa Poikiloblastic amphibolite	q Quartzite
db Diabase	mdb Meta-diabase	ca Chlorite amphibolite	J Jade
Quartz veins	mg Meta-gabbro	cas Chlorite-amphibolite schist	Vertical mine shaft
p Pegmatite	apa Amphibole-plagioclase amphibolite	aqcg Amphibole-quartz-chloritoid gneiss	Mine
			Adit

Figure 19. Seminoe Mountain District (after Bishop, 1964).

Mine Descriptions

King Mine

geology: The King Mine was developed on a five-foot-thick quartz vein. In addition to gold, the vein contains chalcopyrite and pyrite. Seventy tons of ore processed from the mine produced about 36 ounces of gold (Osterwald et al., 1966).

Charlie's Glory Hole

location: Sec. 29, T.26N., R.85W.

geology: In 1962, a small prospect pit was developed on a quartz vein in a fault zone carrying free gold. The vein is enclosed by chlorite-muscovite schist (Bishop, 1964).

Star and Hope Mines

geology: The mines are developed on a $4\frac{1}{2}$-foot vein which dips 15 degrees west-northwest at the surface. At about 80 feet depth the vein dips more steeply (55 degrees northwest). The vein contains free-milling gold, auriferous pyrite, and chalcopyrite (Osterwald et al., 1966).

Sunday Morning Mine

location: $SW\frac{1}{4}$ sec. 29, T.26N., R.85W.

geology: Two shafts were driven into white to rusty massive, southeast-trending quartz veins. A single vein bifurcates at the northwest end, and is locally characterized by limonite-filled boxworks. Free gold and chalcopyrite were observed on the mine dumps (Bishop, 1964).

Long Creek Area

location: Sec. 26, T.26N., R.85W.

geology: Abundant copper minerals occur on a mine dump of a caved shaft driven into a fault in altered amphibole schist. X-ray flourescence analysis of samples from this prospect shows traces of lead, bismuth, gold, silver, uranium, and iron, as well as copper (Bishop, 1964).

BLACK BUTTES DISTRICT

Introduction

The Black Buttes District is located nearly 12 miles west of Negro Hill and approximately eight miles south of Sundance (Figure 20). Samples from this district yielded only traces of gold; but were rich in silver, lead, and zinc.

Economic Geology

Mineralization at Black Butte is found cementing brecciated Paleozoic limestone and as replacement bodies in limestone. The mineralization is localized at the contact between limestone and intrusive andesite porphyry. The ore mineralization is erratic, pockety, and unpredictable. An assay of the deposit gave: gold, 0.002 ounces per ton; silver, 2.0 ounces per ton; lead, 13.9 percent by weight; copper, 0.02 percent; iron, 6.5 percent; manganese, 0.3 percent; sulfur, 4.0 percent; and zinc, 5.7 percent (Osterwald et al., 1966).

Apparently, some galena in the Black Buttes area assayed 100 to 200 ounces of silver, but these rich ores were very restricted (Knight, 1893).

THE LOST CABIN GOLD MINE

Essentially every state in the West has a legend about a lost gold mine, in which the discoverers of fabulous riches of the precious metal were either killed or could not find their way back to their original point of discovery. Wyoming is no exception.

The Lost Cabin gold mine, according to historical records, was never really developed into a mine, but was a placer deposit located in a park (treeless area) along a creek in the Bighorn or Bridger mountains of north central Wyoming. Reports claim that seven prospectors made the discovery in the Fall of 1865 and worked the property for three to six days before they were set upon by Indians who killed five of the seven men. The placer was reported to be so rich that gold could be picked up off the surface with little effort. Two men escaped with an estimated 350 ounces of coarse gold. The only monument to identify the discovery was a small log cabin, a flume, and at least one prospect pit three to four feet deep.

The survivors traveled only at night for three nights on foot until they reached Fort Reno. In the Spring, they hired 8 to 10 prospectors to return to their gold discovery, and the entire party was wiped out by Indians (The Mining Record, Dec. 26, 1979, p. 4). Several people have examined the Bighorn and Bridger mountains looking for the old Lost Cabin Gold mine, and to this day it remains a legend.

THE SIERRA MADRE MINE DISTRICTS

The Sierra Madre Mountains in southeastern Wyoming contain a considerable number of mines and prospects (Plate 1). The majority of the Sierra Madre mines were developed on copper prospects. Although very few gold mines are located in the Sierra Madres, a significant amount of gold must have been produced as a by-product of copper milling. Potential host rocks for gold deposits, in addition to metallic sulfides, include Precambrian quartz-pebble conglomerates (Graff, 1979; Houston et al., 1978); shear zone tectonites along the Nash Fork - Mullen Creek shear zone; mafic igneous rocks (Houston et al., 1975); and the Green Mountain Formation, which is a series of Proterozoic calc-alkalic metavolcanic and volcanogenic metasedimentary rocks similar to Canadian rock with affinities to the volcanogenic massive sulfide deposits (see, e.g., Divis, 1976; 1977).

SUMMARY

Although Wyoming is not known to have been a major gold producing state like many of its neighbors (i.e. South Dakota, Nevada, Utah, Arizona, Colorado, Alaska, California, Montana, and Idaho), several geological environments within the state suggest the potential for the presence of economical base and precious metal deposits. These environments include (1) several placer alluvial deposits formed by the reworking of older sediments; (2) Tertiary porphyry copper-molybdenum deposits; (3) Precambrian quartz pebble conglomerates; and (4) volcanogenic metasedimentary environments.

Continued examination and exploration of Wyoming's potential economic environments may lead to the discovery of economic mineralization. However, federal government programs designed to preserve the environment of many of the State's mountainous regions and basins will undoubtedly limit potential base and precious metal production in the future.

REFERENCES

Antweiler, J. C., and Love, J. D. 1967, Gold-bearing sedimentary rocks in Northwest Wyoming - a preliminary report: U. S. Geol. Survey Circular 541, 12 p.

Antweiler, J. C., Love, J. D., and Campbell, W. L., 1977, Gold content of the Pass Peak Formation and other rocks in the Rocky Mountain Overthrust Belt, Northwestern Wyoming: Wyo. Geol. Assoc., 29th Annual Field Conf. Guidebook, p. 731-749.

Armstrong, F. C., 1947, Preliminary report on the geology of the Atlantic City - South Pass mining district, Wyoming: U.S. Geol. Survey open-file report, Wyo. Geol. Survey files, 64 p.

Bane, J. R., 1929, Report on B and H Mining Company: Unpublished report in Wyo. Geol. Survey files, 10 p.

Bartlett, A. B., and Runner, J. J., 1926, Atlantic City - South Pass gold mining district: Wyo. Geol. Survey Bull. 20, 23 p.

Bayley, R. W., 1965, Geologic map of the Atlantic City quadrangle, Fremont County, Wyoming: U. S. Geol. Survey Map GQ - 459.

Bayley, R. W., 1965, Geologic map of the Louis Lake quadrangle, Fremont County, Wyoming: U. S. Geol. Survey Map GQ - 461.

Bayley, R. W., 1965, Geologic map of the Miners Delight quadrangle, Fremont County, Wyoming: U. S. Geol. Survey Map GQ - 460.

Bayley, R. W., 1965, Geologic map of the South Pass quadrangle, Fremont County, Wyoming: U. S. Geol. Survey Map GQ - 458.

Bayley, R. W., Proctor, P. D., and Condie, K. C., 1973, Geology of the South Pass Area, Fremont County, Wyoming: U. S. Geol. Survey Prof. Paper 793, 39 p.

Beeler, H. C., 1901, Report on the Home, Fayand, and Minnehaha placers: Unpublished report no. 10 for Wyo. Geol. Survey, 4 p.

Beeler, H. C., 1904, Report on the Maudem Group, Lake Creek near Holmes, Albany County, Wyoming: Unpublished report no. 43 for Wyo. Geol. Survey, 3 p.

Beeler, H. C., 1905, Report on the Gold Crater Group, Keystone, Albany County, Wyoming: Unpublished report no. 64 for the Wyo. Geol. Survey, 6 p.

Beeler, H. C., 1906, Mineral and allied resources of Albany County, Wyoming: Laramie, Wyo., The Republican Press, 80 p.

Beeler, H. C., 1907, A brief report on the Kansas Group near Keystone, Albany County, Wyoming: Unpublished report no. 95 for the Wyo. Geol. Survey, 3 p.

Beeler, H. C., 1908, A brief review on the South Pass gold district Fremont County, Wyoming: Wyo. Geol. Survey files, 24 p.

Bishop, D. T., 1964, Retrogressive metamorphism in the Seminoe Mountains, Carbon County, Wyoming: Unpublished M.S. thesis, Univ. of Wyoming, 49 p.

Bolmer, R. I., and Biggs, P., 1965, Mineral resources and their potential on Indian lands, Wind River Reservation, Fremont and Hot Springs counties, Wyoming: U. S. Bur. of Mines, Prel. Rpt. 159.

Childers, M. O., 1957, Geology of the French Creek area, Albany and Carbon counties, Wyoming: Unpublished M.S. thesis, Univ. of Wyo., 58 p.

Curran, H. T., 1926, A general report on the Carissa mining property owned by the Federal Gold Mining Company: Unpublished report in Wyo. Geol. Survey files, 9 p.

Curry, D. R., 1965, The Keystone gold-copper prospect area, Albany County, Wyoming: Wyo. Geol. Survey, Prel. Rpt. no. 3, 12 p.

Dart, A. C., 1929, Report on the Utopia tunnels of the Utopia Mining and Milling Company: Wyo. Geol. Survey files, 10 p.

DeLuguna, W., 1938, Geology of the Atlantic City District, Wyoming: Unpublished Ph.D Thesis, Harvard Univ., 218 p.

Divis, A. F., 1976, The geology and geochemistry of the Sierra Madre Mountains, Wyoming: Colorado School of Mines Quart. v. 71, no. 3, p. 18-21

Divis, A. F., 1977, Isotopic studies on a Precambrian geochronologic boundary, Sierra Madre Mountains, Wyo: Geol. Soc. Amer. Bull., v. 88, p. 96-100.

Dreier, J. E., 1967, Economic geology of the Sunlight mineralized region, Park County, Wyoming: Unpublished M.S. thesis, Univ. of Wyoming., 81 p.

Elliott, J. E., 1980, Mineralization of the Sunlight Mining Region, *in* Geology and mineral resources of the North Absaroka Wilderness and vicinity, Park County, Wyoming: U. S. Geol. Survey Bull. 1447, p. 51-53.

Fisher, F. S., 1972, Tertiary mineralization and hydrothermal alteration in the Stinkingwater mining region, Park County, Wyoming: U. S. Geol. Survey Bull. 1332-C, 33 p.

Fisher, F. S., Antweiler, J. C., and Welsch, E. P., 1977, Preliminary geological and geochemical results from the Silver Creek and Yellow Ridge mineralized areas in the Washakie Wilderness, Wyoming: U. S. Geol. Survey open-file rpt. 77-225, 11 p.

Graff, P. J., 1979, A review of the stratigraphy and uranium potential of early Proterozoic (Precambrian X) metasediments in the Sierra Madre, Wyoming: Univ. of Wyoming Contr. to Geol., v. 17, no. 2, p. 149-157.

Greene, E. A., 1896, Oregon Butte placer mines, Fremont County, Wyoming: Unpub. rept., Univ. Wyoming Archives, Laramie, Wyo., 27 p.

Haff, J. C., 1944, Big Nugget gold placer claims: Unpublished report for Wyo. Geol. Survey, 2 p.

Hausel, W. D., and Glass, G. B., 1980 Mineral resources map for Natrona County: Wyo. Geol. Survey County Resource Series no. 6.

Hausel, W. D., and Holden, G. S., 1978, Mineral resources of the Wind River Basin and adjacent Precambrian uplifts: Wyo. Geol. Assoc., 30th Field Conference Guidebook, p. 303-310.

Hess, F. L., 1926, Platinum near Centennial, Wyoming: U. S. Geol. Surv. Bull. 780-C, p. 127-135.

Houston, R. S., and others, 1968, A regional study of rocks of Precambrian age in that part of the Medicine Bow Mountains lying in southeastern Wyoming - with a chapter on the relationship between Precambrian and Laramide structure: Wyo. Geol. Survey Memoir no. 1, 167 p.

Houston, R. S., and others, 1975, Preliminary report on distribution of copper and platinum group metals in mafic igneous rocks of the Sierra Madre, Wyoming: Wyo. Geol. Survey, open-file rpt. 75-85, 129 p.

Houston, R. S., Karlstrom, K. E., Graff, P. J., and Hausel, W. D., 1978, Radioactive quartz pebble conglomerates of the Sierra Madre and Medicine Bow Mountains, southeastern Wyoming: Wyo. Geol. Survey open-file rpt. 78-3, 49 p.

Jamison, C. E., 1911, Geology and mineral resources of a portion of Fremont County: Wyo. Geol. Survey Bull. no. 2, 90 p.

Keeton, C, G., 1913, Report for the Mother Lode Gold and Copper Company: Unpublished report in the Wyo. Geol. Survey files, 3 p.

King, J. R., 1961, Geology of the Boswell Creek Area, Albany County, Wyoming: Unpublished M.A. thesis, Univ. of Wyo., 83 p.

Knight, W. C., 1893, Notes on the mineral resources of the State: Univ. of Wyo. Exper. Station Bull. 14, p. 103-212.

Knight, S. H., 1942, Known mineral resources of Albany County, Wyoming: Prepared for Albany County Council of Defense, Wyo. Geol. Survey files, 24 p.

Koschman, A. H., and Bergendahl, M. H., 1968, Principal gold producing districts of the United States: U. S. Geol. Survey Prof. Paper 610, 283 p.

Kyner, T., 1907, Report on the geology of Miner's Delight Group: Unpublished report in Wyo. Geol. Survey files, 10 p.

Lindsey, D. A., 1972, Sedimentary petrology and paleocurrents of the Harebell Formation, Pinyon Conglomerate, and associated coarse clastic deposits, northwestern Wyoming: U. S. Geol. Survey Prof. Paper 734-B, 68 p.

Love, J. D., and others, 1955, Geologic map of Wyoming (1:500,000): U.S. Geol. Survey.

Love, J. D., 1973, Harebell Formation (Upper Cretaceous) and Pinyon Conglomerate (Uppermost Cretaceous and Paleocene), northwestern Wyoming: U. S. Geol. Survey Prof. Paper 734-A, 54 p.

Love, J. D., Antweiler, J. C., and Mosier, E. L., 1978, A new look at the origin and volume of the Dickie Springs - Oregon Gulch placer gold at the south end of the Wind River Mountains: Wyoming Geological Association, 13th Annual Field Conference Guidebook, p. 379-391.

Michalek, D. D., 1952, Precambrian geology of Jelm Mountain, Albany County, Wyoming: Unpublished M.A. thesis, Univ. of Wyo., 51 p.

McCallum, M. E., 1968, The Centennial Ridge gold-platinum district, Albany County, Wyoming: Wyo. Geol. Survey Prel. Rpt. no. 7, 13 p.

McCallum, M. E., and Orback, C. J., 1968, The New Rambler copper-gold-platinum district, Albany and Carbon counties, Wyoming: Wyo. Geol. Survey Prel. Rpt. no. 8, 12 p.

Molker, A. J., 1923, History of Natrona County: R. R. Donnelley and Sons Co., Chicago, The Lakeside Press, p. 96-102.

Osterwald, F. W., Osterwald, D. B., Long, J. S., Jr., and Wilson, W. H., 1966, Mineral resources of Wyoming: Wyo. Geol. Surv. Bull. 50, 287 p. (Revised by W. H. Wilson).

Parsons, W. H., 1937, The ore deposits of the Sunlight mining region, Park County, Wyoming: Econ. Geol., v. 32, p. 832-854.

Peterson, K., 1968, Sulfide mineralization of the northwest part - Sunlight mining region, Park Co., Wyoming: Unpublished M.S. thesis, Univ. of Wyoming, 49 p.

Phillips, A., 1907, Report to the Federal Gold Mining Company: Unpublished report in Wyo. Geol. Survey files, 22 p.

Prinz, W. C., 1974, Map showing geochemical data for the Atlantic City gold district, Fremont County, Wyoming: U. S. Geol. Survey Map I-865.

Ramond, R. W., 1870, Statistics of mines and mining in the States and Territories west of the Rocky Mountains: U. S. Dept. of Treasury, 805 p.

Rich, D. H., 1974, Economic geology of the Silvertip Basin, Sunlight mining region, Park County, Wyoming: Unpublished M.S. thesis, Miami Univ., 73 p.

Rosenkranz, R. D., Davidoff, R. L., and Lemons, J. F., Jr., 1979, Copper availability - domestic: U. S. Bureau of Mines IC-8809, 31 p.

Schoen, R., 1953, Geology of the Cooper Hill District, Carbon County, Wyoming: Unpublished M.A. thesis, Univ. of Wyo., 41 p.

Schultz, A. R., 1907, Gold developments in central Uinta County, Wyoming, and at other points on Snake River: U. S. Geol. Survey Bull. 315A, p. 71-88.

Spencer, A. C., 1916, The Atlantic City gold district and the North Laramie Mountains: U. S. Geol. Survey Bull. 626, 85 p.

Staley, W. W., 1946, Gold in Idaho: Idaho Bur. Mines and Geology, Pamphlet no. 68.

Trumbull, L. W., 1914, Atlantic City gold mining district, Fremont County: Wyo. Geol. Survey Bull. no. 7, 100 p.

Williams, F. E., 1980, Metallic mineral deposits, *in* Geology and mineral resources of the North Absaroka Wilderness and vicinity, Park County, Wyoming: U. S. Geol. Survey Bull. 1447, p. 33-51.

Wilson, W. H., 1951, Report on the Atlantic City - South Pass mining district: Unpublished report for Wyo. Geol. Survey, 1 p.

Wilson, W. H., 1953, Notes on the Wyoming Mica and Metals Corporation, Lander, Fremont County, Wyoming: Unpublished report for Wyo. Geol. Survey, 2 p.

Wilson, W. H., 1960, Petrology of the Wood River area, southern Absaroka Mountains, Park County, Wyoming: Unpublished Ph.D thesis, Univ. of Utah, 121 p.

Wilson, W. H., 1964, The Kirwin mineralized area, Park County, Wyoming: Wyoming Geol. Survey Prel. Rpt. no. 2, 12 p.

Wilson, W. H., 1975, The Copper-bearing Meadow Creek granodiorite, upper Wood River area, Park County, Wyoming: Wyo. Geol. Assoc., 27th Ann. Field Conf. Guidebook, p. 235-242.

Wyoming State Mine Inspector, Annual report for 1957, p. 49.

Zeller, H. D., and Stephens, E. V., 1964, Geologic map of the Dickie Springs quadrangle Fremont and Sweetwater counties, Wyoming: U. S. Geol. Survey Map MF-293.

Zeller, H. D., and Stephens, E. V., 1969, Geology of the Oregon Butte area, Sweetwater, Sublette, and Fremont counties, southwestern Wyoming: U. S. Geol. Survey Bull. 1256, 60 p.

www.ingramcontent.com/pod-product-compliance
Lightning Source LLC
Chambersburg PA
CBHW051418200326
41520CB00023B/7277